# NATIONAL 4 & 5

# GEOGRAPHY

## GLOBAL ISSUES

### SECOND EDITION

Calvin Clarke & Susie Clarke

The Publishers would like to thank the following for permission to reproduce copyright material:

**Photo credits**

Chapter opener image on **pp.1, 7, 13, 19, 22, 26** and **32** © tusharkoley – Fotolia; **p.5 (top left)** © S. Wanke/Photodisc/Getty Images, **(top right)** © ARCTIC IMAGES / Alamy Stock Photo, **(bottom left)** © tusharkoley – Fotolia; **p.6 (left to right)** © Siede Preis/Photodisc/Getty Images, © Siede Preis/Photodisc/Getty Images, © mirpic – Fotolia; **p.9** NASA/Photodisc/Getty Images/ Environmental Concerns 31; **p.15** © V. ZHURAVLEV – Fotolia; **p.17 (top left)** © Mikael Damkier / Fotolia.com, **(top right)** © Elwynn - Fotolia.com, **(centre left)** © Elwynn - Fotolia.com, **(centre right)** © Lubos Chlubny – Fotolia, **(bottom left)** © david hughes – Fotolia, **(bottom right)** © Sergiy Serdyuk – Fotolia; **p.18 (top left)** © eag1e – Fotolia, **(top right)** © TMAX – Fotolia, **(bottom left)** © corepics - Fotolia.com; **p.27** © jjayo - Fotolia.com; **p.33** © Getty Images/ Stockbyte Platinum/Thinkstock; **p.35** © Pat Canova / Alamy; Chapter opener image on **pp.44, 47, 52, 58, 64, 67, 74, 79, 85, 90** and **96** NASA/Photodisc/Getty Images/ Environmental Concerns 31; **p.60** U.S. Geological Survey; **p.62** © CORBIS/Corbis via Getty Images; **p.65 (top left)** U.S. Geological Survey/photo by D. Dzurisin. Cowlitz County, Washington. May 31, 1980, **(top right)** © Bart Rayniak/The Spokesman-Review, **(bottom right)** © David R. Frazier Photolibrary, Inc. / Alamy; **p.66** © Gary Braasch/Corbis Documentary/Getty; **p.80 (top)** © AlamyCelebrity / Alamy, **(bottom)** © WENN UK / Alamy Stock Photo; **p.95 (top left)** © U.S. Navy via Getty Images, **(top right)** © Mark Wolfe/FEMA, **(bottom)** © Michael Rieger/FEMA; **p.97** © AP/REX/Shutterstock; Chapter opener image on **pp.106, 111, 115, 121** © Incredible Arctic - stock.adobe.com; **p.108** © Design Pics Inc/REX/Shutterstock; **p.109** © Alexander Piragis - stock.adobe.com; **p.112 (top)** © Credit: National Geographic Creative / Alamy Stock Photo, **(bottom)** © Credit: All Canada Photos / Alamy Stock Photo; **p.116** © Shutterstock / Gary Whitton; **p.117** © Credit: Design Pics Inc / Alamy Stock Photo; **p.118** © corepics - stock.adobe.com; **p.120** © Credit: Accent Alaska.com / Alamy Stock Photo; **p.122 (top)** © Credit: Danita Delimont / Alamy Stock Photo, **(bottom)** © magele-picture - stock.adobe.com; Chapter opener image on **pp. 125, 131, 136** and **140** © Olivier Rault - stock.adobe.com; **p.127** © cn0ra - stock.adobe.com; **p.128** © Juhku - stock.adobe.com; **p.133** © Credit: Westend61 GmbH / Alamy Stock Photo; **p.135 (top left)** © ANTONIO SCORZA/AFP/Getty Images, **(top right)** © whitcomberd - stock.adobe.com, **(centre left)** © Credit: BrazilPhotos.com / Alamy Stock Photo, **(centre right)** © Credit: Sue Cunningham Photographic / Alamy Stock Photo, **(bottom)** © Andre Penner/AP/REX/Shutterstock; **p.137 (top)** © Richard Carey - stock.adobe.com, **(bottom)** © filipefrazao - stock.adobe.com; **p.141 (top)** © Arco Images GmbH / Alamy Stock Photo, **(bottom)** © Surrey Satellite Technology Ltd; Chapter opener image on **pp.148, 153, 158, 162, 166, 172, 177** and **180** © punyafamily – Fotolia; **p.156** © Iamnee – Fotolia; **p.157 (top)** Image Courtesy of The Advertising Archives, **(bottom)** © REUTERS/Will Burgess; **p.159** © Henrik Larsson – Fotolia; **p.164** © punyafamily – Fotolia.

**Acknowledgements**

The text extracts on pages iv–v of the Introduction have been quoted from pages 4 and 8 of the National 5 Geography Course Specification, and have been reproduced with the permission of the Scottish Qualifications Authority. Copyright © Scottish Qualifications Authority.

Every effort has been made to trace all copyright holders, but if any have been inadvertently overlooked, the Publishers will be pleased to make the necessary arrangements at the first opportunity.

Although every effort has been made to ensure that website addresses are correct at time of going to press, Hodder Gibson cannot be held responsible for the content of any website mentioned in this book. It is sometimes possible to find a relocated web page by typing in the address of the home page for a website in the URL window of your browser.

Hachette UK's policy is to use papers that are natural, renewable and recyclable products and made from wood grown in sustainable forests. The logging and manufacturing processes are expected to conform to the environmental regulations of the country of origin.

Orders: please contact Bookpoint Ltd, 130 Park Drive, Milton Park, Abingdon, Oxon OX14 4SE. Telephone: (44) 01235 827827. Fax: (44) 01235 400401. Email education@bookpoint.co.uk Lines are open from 9 a.m. to 5 p.m., Monday to Saturday, with a 24-hour message answering service. Visit our website at www.hoddereducation.co.uk. Hodder Gibson can also be contacted at hoddergibson@hodder.co.uk

© Calvin Clarke and Susie Clarke 2018

First published in 2013 © Calvin Clarke and Susie Clarke

This second editions published in 2018 by

Hodder Gibson, an imprint of Hodder Education

An Hachette UK Company

211 St Vincent Street

Glasgow, G2 5QY

| Impression number | 5 | 4 | 3 | 2 | 1 |
|---|---|---|---|---|---|
| Year | 2022 | 2021 | 2020 | 2019 | 2018 |

All rights reserved. Apart from any use permitted under UK copyright law, no part of this publication may be reproduced or transmitted in any form or by any means, electronic or mechanical, including photocopying and recording, or held within any information storage and retrieval system, without permission in writing from the publisher or under licence from the Copyright Licensing Agency Limited. Further details of such licences (for reprographic reproduction) may be obtained from the Copyright Licensing Agency Limited, www.cla.co.uk

Cover photo © Shutterstock / siriwat sriphojaroen

Illustrations by Peters & Zabransky Ltd and Integra Software Services Pvt. Ltd., Pondicherry, India

Typeset in 11 on 12 pt Stempel Schneidler Std Light by Integra Software Services Pvt. Ltd., Pondicherry, India

Printed in Slovenia

A catalogue record for this title is available from the British Library.

ISBN: 978 1 5104 2938 3

# Contents

|     | Introduction | iv |
| --- | --- | --- |
| 1. | Climate change | 1 |
| 2. | Climate change – physical factors | 7 |
| 3. | Climate change – human factors | 13 |
| 4. | Climate change – its effects | 19 |
| 5. | Climate change – coping with its effects | 22 |
| 6. | Climate change – case study of Bangladesh | 26 |
| 7. | Climate change – case study of Florida, USA | 32 |
| 8. | Structure of the Earth | 44 |
| 9. | Crustal plates and plate boundaries | 47 |
| 10. | Volcanoes | 52 |
| 11. | The eruption of Mt. St. Helens, 1980 | 58 |
| 12. | Managing the eruption of Mt. St. Helens, 1980 | 64 |
| 13. | Earthquakes | 67 |
| 14. | The cause of the Japan earthquake, 2011 | 74 |
| 15. | The effects and management of the Japan earthquake, 2011 | 79 |
| 16. | Tropical storms | 85 |
| 17. | The cause of Hurricane Irma, 2017 | 90 |
| 18. | The impact and management of Hurricane Irma, 2017 | 96 |
| 19. | The tundra climate | 106 |
| 20. | How the tundra environment is used and misused | 111 |
| 21. | Alaska: the effects of human activities | 115 |
| 22. | The management of human activities in the tundra | 121 |
| 23. | The equatorial rainforest climate | 125 |
| 24. | Uses of the equatorial rainforest | 131 |
| 25. | The Amazon rainforest: effects of deforestation | 136 |
| 26. | The management of human activities in equatorial rainforests | 140 |
| 27. | Health in developing countries | 148 |
| 28. | Health in developed countries | 153 |
| 29. | Malaria – its cause and transmission | 158 |
| 30. | Malaria – factors in its distribution | 162 |
| 31. | Heart disease – its causes | 166 |
| 32. | Heart disease – methods of control | 172 |
| 33. | HIV/AIDS – its distribution | 177 |
| 34. | HIV/AIDS – causes, effects and treatment | 180 |
|     | 'I can do' | 189 |
|     | Index | 194 |

# Introduction

This book has been written to cover one of the three areas of study in the Scottish Qualification Authority's National 4 and National 5 courses, Global Issues.

## Global Issues

The Course Specification for National 4 and National 5 states:

*Candidates develop skills in using numerical information in the context of global issues, together with a detailed knowledge and understanding of significant global geographical issues. Key topics include: climate change, natural regions, environmental hazards, trade and globalisation, tourism, and health.*

Of the six key topics, you should study two. Four are covered in this book. As outlined below, specific case studies are provided for each topic, to illustrate and support students' understanding of the issues covered. Candidates should note, however, that in the exam, they can refer to other case studies and examples in their responses.

### Climate change

This topic is covered in Chapters 1–7.

- Features of climate change
- Causes – physical and human
- Effects – local and global
- Management strategies to minimise impact/effects.

Case studies of Florida and Bangladesh are given to help candidates understand the effects of climate change and the strategies employed to reduce its impact.

### Environmental hazards

This topic is covered in Chapters 8–18.

- The main features of earthquakes, volcanoes and tropical storms

- Causes of each hazard
- Impacts of each hazard on people and the landscape
- Management – methods of prediction and planning, and strategies adopted in response to environmental hazards.

Case studies of Mt. St. Helen's volcano, the Japan earthquake of 2011 and Hurricane Irma in 2017 are given.

## Natural regions

This topic is covered in Chapters 19–26.

- Tundra and equatorial tropical forest climates and their ecosystems
- Use and misuse of these environments by people
- Effects of land degradation on people and the environment
- Management strategies to minimise impact/effects.

Case studies of the Alaska tundra region and the Amazon rainforest are given.

## Health

This topic is covered in Chapters 27–34.

- Distribution of a range of world diseases
- Causes, effects and strategies adopted to manage:
  - HIV/AIDS in developed and developing countries
  - one disease prevalent in a developed country (choose from: heart disease, cancer, asthma)
  - one disease prevalent in a developing country (choose from: malaria, cholera, kwashiorkor, pneumonia).

In this section, HIV/AIDS, heart disease and malaria are covered.

## N4- and N5-level questions and activities

Each chapter contains N4- and N5-level questions and activities. These are designed to develop your knowledge and understanding of global issues, but also to develop a range of skills. These include literacy, numeracy, enterprise, citizenship and thinking skills. The N4 questions and N5 questions test knowledge and understanding of the course content; the activities test the same concepts but encourage active learning and the development of a wider range of skills.

Answers to questions at N4 and N5 are differentiated chiefly according to the amount of detail given. The activities are differentiated by student response.

At the end of each topic (Chapters 7, 18, 26 and 34) there is an additional set of questions. These give you examples of N4 Assessment questions and N5 SQA-style questions on the topics covered in the preceding chapters. There is also advice on answering these questions and some related tasks.

Each chapter in the book begins by stating the learning intentions. At the end of each chapter you are asked to self-assess your understanding of these learning intentions using the traffic-light system. A photocopiable checklist for all the learning intentions is found at the back of this book for you to use. This 'I can do' self-assessment approach is explained on the following page.

## 'I can do'

Each chapter in this book has a box at the beginning outlining what you will be learning and what you should be able to do after you have completed the N4/N5 questions and/or Activities.

It is very important that you feel confident about these as you will be assessed on them.

After you have completed each chapter, you will be asked to fill out the 'I can do' boxes for that chapter. These can be found at the back of the book from page 189.

The 'I can do' checklist outlines all of the learning intentions for every chapter within this book. You have to fill it out based on how well you understood the information in the chapter. It is a 'traffic light' system:

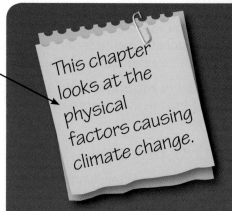

This chapter looks at the physical factors causing climate change.

### By the end of this chapter you should be able to:

✓ describe how physical factors affect the world's climate
✓ give examples of some physical factors that cause climate change
✓ explain how these factors cause climate change.

### Now complete the 'I can do' boxes for this chapter.

- **RED** means that you DO NOT FEEL THAT YOU UNDERSTAND THIS AND DON'T THINK YOU CAN DO THIS NOW.
- **YELLOW** means that you THINK YOU CAN DO MOST OF IT BUT YOU STILL HAVE SOME PROBLEMS.
- **GREEN** means that you FULLY UNDERSTAND THIS AND CAN DO IT WITHOUT ANY DIFFICULTY.

On the 'I can do' checklist, there is also a space for comments. It is worthwhile taking a couple of minutes to write a few comments as you complete each chapter, as they will prove very helpful when you start revising. See the example below.

|  | Red | Yellow | Green | Comment |
|---|---|---|---|---|
| **Chapter 1 Climate change** | | | | |
| Describe what climate change is | | ✓ | | I understand what climate change is but I'm still unsure how to describe it concisely. |
| Describe the natural greenhouse effect | | | ✓ | I understand what the natural greenhouse effect is and can describe it accurately. |
| Give examples of evidence that our climate has changed | | | ✓ | I can give detailed examples of evidence that our climate has changed. |

# Chapter 1

# Climate change

This chapter looks at how the world's climate is changing.

**By the end of this chapter, you should be able to:**

- ✓ describe what climate change is
- ✓ describe the natural greenhouse effect
- ✓ give examples of evidence that our climate has changed.

### Did you know...?
There are many different greenhouse gases, such as water vapour, carbon dioxide, methane, nitrous oxide and ozone.

## The natural greenhouse effect

The Earth is surrounded by an invisible layer of air called the atmosphere. In the atmosphere there are many different gases, chiefly nitrogen and oxygen but also several others in much smaller amounts. They are all important to us in different ways. **Only some of the gases in the atmosphere are greenhouse gases**; but they are essential because they help to keep the Earth warm. Without the presence of these greenhouse gases, life on Earth could not exist. Figure 1.1 illustrates this natural greenhouse effect.

The Earth is heated by the sun. It emits heat energy that passes through the atmosphere and heats the Earth. When the Earth heats up, it also emits heat energy. But some of this heat can now be absorbed by the greenhouse gases in the atmosphere, instead of escaping back into space. This keeps the Earth warm and is called the natural greenhouse effect. This is because **these gases act just like the panes of glass in a greenhouse; they allow the sun's heat in but they trap some of the heat trying to get out.** So the greenhouse becomes hotter than the air outside.

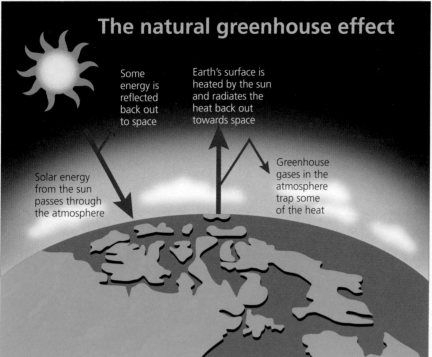

**Figure 1.1**
The natural greenhouse effect

# Climate change

The term **'climate change' means a significant change in global weather patterns over a long period of time**. Figure 1.2 shows how temperatures have changed in the last 100 years. You can see that temperatures have been rising since 1975. Figure 1.3 shows how temperatures have changed in the last 10,000 years. You can see that they have changed even more. At times we have been a little warmer and

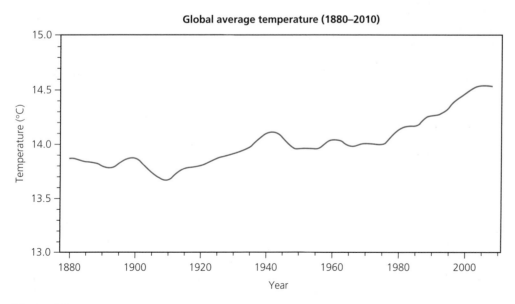

**Figure 1.2**
Temperature change over the last 100 years

# CLIMATE CHANGE

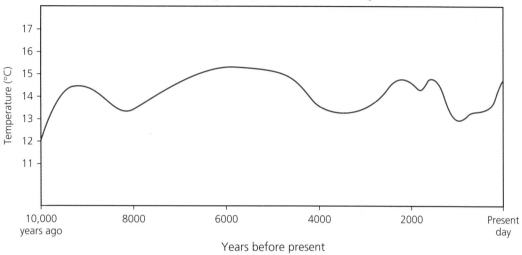

**Figure 1.3**
Temperature change over the last 10,000 years

10,000 years ago much colder. In fact, climate change has been taking place ever since the Earth was formed and it will surely change just as often in the future.

Because climate change has been going on for so long, we know that human beings cannot have been responsible for all of it. So **climate change must be due to both natural causes and human activities**.

Every country in the world has experienced some form of climate change over the last 10,000 years. We know this because records have been collected from a number of different sources, highlighting long-term, medium-term and short-term changes. Table 1.1 below describes some of these changes.

| Long-term changes (over many thousands of years) | Medium-term changes (over many hundreds of years) | Short-term changes (over the last 100–200 years) |
|---|---|---|
| **Ice core analysis** This is a very effective way of collecting long-term records of temperature changes. In polar regions snow falls every year and never melts. Over time, the layers of snow compact under their own weight to form ice. By drilling through that ice and taking samples of it, records of temperature and atmospheric gases can be built up for periods of hundreds of thousands of years. The samples that are taken contain dust, ash, gas and air bubbles and radioactive substances. These can then be analysed to build a climate record. | **Tree ring analysis** Trees tend to make one growth ring each year, with the newest ring nearest the bark. A year-by-year record or ring pattern is formed that reflects the climate conditions in which the tree grew. For example, a wide ring shows that it was warm and wet because the tree had adequate moisture and a long growing season, while a dry or cold year would result in a very narrow ring. | **Recording changes in biodiversity** Different species of animals and plants are adapted to certain environments. If the conditions change then they have three options: move, adapt or die. By recording changes to the numbers and types of animals and plants in different areas, we can tell how the climate is changing there. |

*continued*

| Long-term changes (over many thousands of years) | Medium-term changes (over many hundreds of years) | Short-term changes (over the last 100–200 years) |
|---|---|---|
| | **Glaciers** Glaciers are also a good source of information on climate change. Glaciers are moving rivers of ice and they flow downhill due to gravity. They are very sensitive to temperature fluctuations. If temperatures are consistently low, the glacier will grow and advance; if there is an increase in temperatures the glacier will retreat (melt). It will leave evidence of where it used to be. | **Ice extent** Keeping an annual record of the extent of ice around the world is a good way of showing changes to our climate. |
| | | **Measuring air and sea temperatures** Making a continuous record of the temperature is a good way of showing whether our climate is changing. By recording temperatures all the time we can compare them with previous years. We have been doing this for over 100 years. |

**Table 1.1**
Examples of measuring long-term, medium-term and short-term climate changes

# National 4

1. Using the word bank below, copy and complete the paragraph, which explains how our atmosphere helps to keep the Earth warm.
   The sun emits heat energy that passes through the Earth's _____. The Earth heats up and some of its heat energy is radiated back into the atmosphere. The greenhouse gases in the atmosphere _____ this heat energy and store it in the atmosphere rather than letting it _____ into space. This is called the _____ . The main greenhouse gases are _____, _____ and _____.

   carbon dioxide      natural greenhouse effect      methane      absorb
   atmosphere          water vapour                   escape

2. Draw a similar diagram to Figure 1.1 in your jotter and label it with the following information:
   - Atmosphere
   - Greenhouse gases
   - Arrow showing the sun's radiation
   - Arrow showing heat energy from the Earth

3. What does the term 'climate change' mean?
4. Look at Figure 1.2. What was the average temperature in (a) 1880, (b) 2010?
5. Look at Figure 1.3. What was the average temperature (a) 10,000 years ago, (b) 6000 years ago, (c) 2000 years ago?
6. We have sources of evidence that our climate has changed. Explain how we know that there have been changes to our climate:
   (a) in the short term
   (b) in the medium or long term.

# CLIMATE CHANGE

## National 5

1. Describe, in detail, how the atmosphere helps to keep the Earth warm enough to sustain life.
2. Draw a diagram similar to Figure 1.1 and annotate it in your own words.
3. Give a definition of the term 'climate change'.
4. Look at Figure 1.2. Describe the changes in the world's average temperature since 1880.
5. Look at Figure 1.3. When were the coldest and warmest periods in the last 10,000 years and how much colder and warmer were they?
6. Choose one of each of the sources showing how our climate has changed and describe it in detail.

## Activities

### Activity A

Look at the pictures below and decide which methods of recording climate change are being shown.

A

B

C

## Activities continued...

### Activity B

Look at the pictures below.

(a) Which picture shows a very cold, dry climate? Why?
(b) Which picture shows a climate which has changed a lot? Why?
(c) Which picture shows a warm, wet climate? Why?

A

B

C

### Activity C

Scientists have been keeping a record of average air temperatures and average sea temperatures in country A. The findings for the last ten years are shown below.

Draw a multiple line graph to show the results.

|         | 2008 | 2009 | 2010 | 2011 | 2012 | 2013 | 2014 | 2015 | 2016 | 2017 |
|---------|------|------|------|------|------|------|------|------|------|------|
| Air (°C)| 10   | 12   | 14   | 11   | 11   | 9    | 12   | 15   | 13   | 11   |
| Sea (°C)| 1    | 1.5  | 2    | 1    | 1    | 0    | 1.5  | 2    | 1.5  | 1    |

Now complete the 'I can do' boxes for this chapter.

# Chapter 2

This chapter looks at the physical factors causing climate change.

# Climate change – physical factors

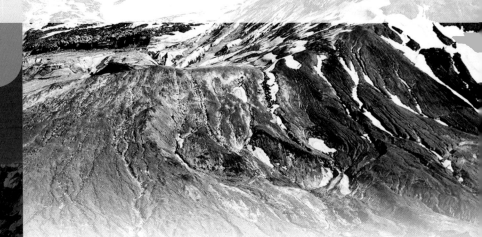

**By the end of this chapter, you should be able to:**

- ✓ describe how physical factors affect the world's climate
- ✓ give examples of some physical factors that cause climate change
- ✓ explain how these factors cause climate change.

**Did you know...?**
The climate has always changed, and it always will.

## Physical factors affecting our climate

Most people instantly think of human activities when trying to explain climate change, not realising that natural events have and have had a huge impact on global climate.

Natural reasons for climate change include changes in solar radiation, changes in oceanic circulation, volcanic eruptions, the Earth's movement and the movement of crustal plates; all of these have played a noticeable part in long-term climate change.

### Changes in solar radiation

**Our climate changes because we do not always receive the same amount of heat energy from the sun.** The number of sunspots on the surface of the sun changes all the time and peaks every eleven years, on average. The greater the number of sunspots, the more heat we receive from the sun. For example, for a period of 70 years starting in the late 1600s there were few sunspots on the sun and our climate became colder (sometimes called 'The Little Ice Age').

## The Earth's stretch, tilt and wobble

The Earth spins on an axis that is tilted at approximately 23° to the vertical and it moves around the sun in an orbit. It takes 24 hours for the Earth to spin on its axis and it takes just over 365 days for it to orbit the sun. **Over thousands of years the Earth's tilt changes, its orbit changes and it wobbles on its axis. All these changes affect our climate.** For example, when the Earth tilts more on its axis and when its orbit takes us closer to the sun, it makes the climate warmer. These events are thought to explain why our Ice Ages ended.

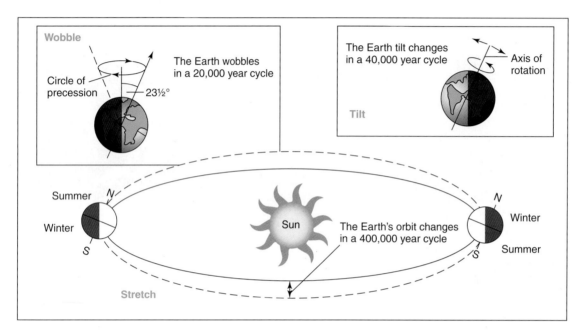

**Figure 2.1**
The Earth's stretch, tilt and wobble

## Changes to ocean currents

The oceans help to transfer heat energy around the world. Within the oceans are ocean currents; some of these are flowing towards the tropics and some are flowing towards the poles. The ocean currents flowing from the tropics to the poles transfer warm water, and vice versa; this affects the climates of countries all over the world. Figure 2.2 shows the main currents that are responsible for transferring this heat energy. In the UK we benefit from the North Atlantic Drift, an ocean current in the Atlantic Ocean that brings warm water from the tropics. It especially affects the west coast and makes our winters 5°C warmer.

**Over time, ocean currents change direction and they can also become stronger or weaker. When they do this they change the climates of the places they flow past.** The North Atlantic Drift suddenly stopped 11,000 years ago, which triggered a short return of the Ice Age (see Figure 1.3). There is also some evidence that it has been slowing down in the last 50 years.

## Volcanic eruptions

During a volcanic eruption lava, ash and gases are ejected from the volcano and sent into the atmosphere. In the atmosphere these particles absorb some of the heat energy from the sun, preventing it from reaching the Earth's surface. This lowers temperatures. There have been many eruptions over the years that have decreased global temperatures, including Mount St. Helens (1980), Mount Hekla (1980) and Mount Pinatubo (1991). In 1815 Mount Tambora in Indonesia erupted and sent more gas and ash into the atmosphere than has any other volcano. The Earth cooled down, Europe received snow in August, crops failed and the following year was known as the year without summer. If one eruption can do this, scientists believe that **a period with a lot of volcanic activity makes the climate colder**.

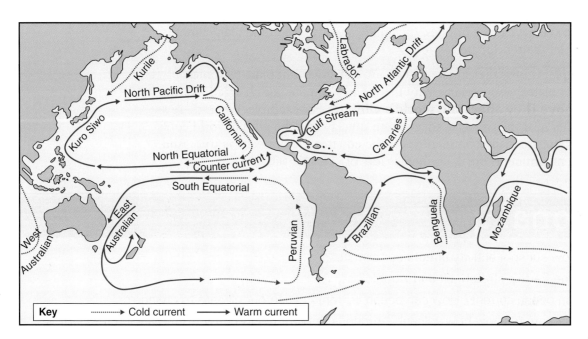

**Figure 2.2**
Ocean currents affect our climates

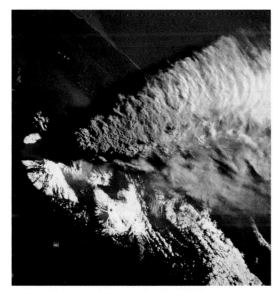

**Figure 2.3**
Volcanic eruptions lower temperatures

## Plate tectonics

The further you are from the equator, the cooler the summer and winter temperatures are likely to be. However, the location of countries today is not where they have always been. Every country, continent and ocean is sitting on top of huge tectonic plates. These plates are constantly moving and have been doing so for hundreds of millions of years.

If we take Britain as an example, approximately 300 million years ago Britain lay on the equator and we had hot, wet weather. Over the years, we have drifted to our current latitude, passing through other climate regions, and so we have had a hot desert climate, a warm Mediterranean climate and now a cooler wetter climate.

This clearly explains why the climate of a country changes but plate tectonics (moving crustal plates) also explains why the world's climate has changed. **As the continents move they affect the world's climate.** For example, when there are many continents near the poles ice sheets form quickly and the whole world cools down; when there are many continents near the equator the opposite is true. And, of course, when continents break up there is lots of volcanic activity.

**Did you know...?** On average, tectonic plates move approximately 5–10 cm each year.

# National 4

1. Explain how sunspot activity affects our climate.
2. What is the Earth's stretch, tilt and wobble?
3. What is an ocean current? How can ocean currents cause our climate to change?
4. Draw a diagram to show how volcanic eruptions prevent heat energy from reaching the Earth's surface. Label it with the following information:
   - Ash and gases
   - Atmosphere
   - Incoming solar energy
   - Earth's surface
5. Describe how the movement of crustal plates can cause climate change.
6. The text mentions three things that have either started or stopped an Ice Age. Describe one of these.

# National 5

1. The Earth does not always receive the same amount of heat from the sun. Explain how this happens.
2. Rearrange the following sentences so that they explain how ocean currents affect climate change.
   - A change in the pattern and strength of ocean currents changes the distribution of heat around the planet.
   - Warm ocean currents transfer warm water from the tropics to the poles.
   - This causes climates to change.

# CLIMATE CHANGE – PHYSICAL FACTORS

## National 5 continued...

- Cold ocean currents replace this warm water by transferring cold water from the poles to the tropics.

3. Describe, in detail, how plate tectonics affects the climates of countries.

*Everybody knows volcanoes are hot when they erupt – very, very hot! How can you say that volcanic eruptions make the climate colder?!*

4. Explain to this person why he is wrong. Use a diagram to help.
5. The text mentions three things that have either started or stopped an Ice Age. Describe two of these.

## Activities

| Physical (natural event) | Hotter or colder climate? | Short-term or long-term change? |
|---|---|---|
| The Earth's tilt on its axis increases to 25° | | |
| The North Pacific Drift suddenly strengthens | | |
| The Earth's orbit takes it further from the sun | | |
| All the continents move nearer the equator | | |
| The supervolcano at Yellowstone erupts | | |
| The Brazilian and Mozambique ocean currents stop flowing | | |
| The African crustal plate breaks apart | | |
| Many sunspots form on the surface of the sun | | |

## Activities continued...

### Activity A

Look at the events in the table above. For each one, decide (a) whether it will make the Earth colder or warmer and (b) whether it will be a short-term change (take place over a few years) or a long-term change (take place over thousands of years). Then draw the table and complete it.

### Activity B

Choose one of the factors that cause climate change and draw a poster to show exactly how it does so.

### Activity C

Using Figure 2.2 and an atlas answer the following questions:

(a) Which ocean current moves warm water from the Gulf of Mexico to Western Europe?
(b) Name one country affected by the Benguela current.
(c) Is the Canaries current warm or cold?
(d) Name two countries in West Africa that are affected by the Canaries current.
(e) Which current passes by Chile?
(f) Which current passes by California?

**Now complete the 'I can do' boxes for this chapter.**

# Chapter 3

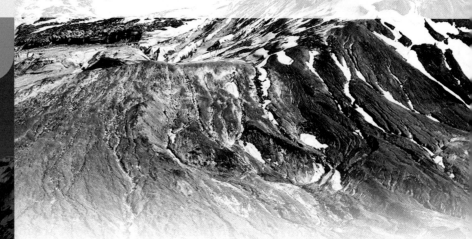

## Climate change – human factors

This chapter looks at the human factors causing climate change.

**By the end of this chapter, you should be able to:**

- ✓ describe the difference between the natural and the enhanced greenhouse effect
- ✓ give examples of some human activities that contribute to climate change
- ✓ explain how these activities affect the climate.

### Did you know...?
Carbon dioxide can stay in the atmosphere for 200 years! Nitrous oxide will persist for about 114 years. Methane remains there for about 12 years.

## The natural and the enhanced greenhouse effect

In Chapter 1 we looked at the importance of the natural greenhouse effect. This is the effect of having small amounts of greenhouse gases in the atmosphere which allow life on Earth to exist because they trap heat and keep us warm. However, if these gases increase in the atmosphere, they can cause global temperatures to increase.

**The enhanced greenhouse effect is caused when human activities release greenhouse gases into the atmosphere** (Figure 3.1). When greenhouse gases are in the atmosphere they absorb heat energy, trapping it and causing temperatures to increase. Greenhouse gases can remain in the atmosphere for many years.

## Human causes of climate change

Most scientists agree that before the Industrial Revolution people did not affect the climate. Any changes to the climate

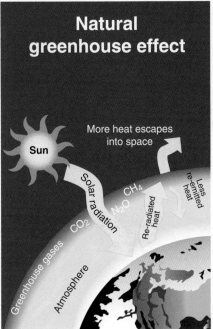
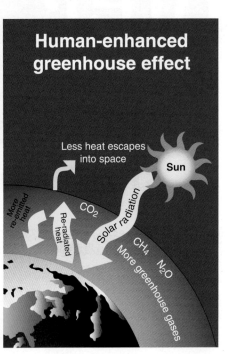
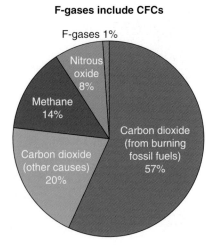

Figure 3.2
Global greenhouse gas emissions

Figure 3.1
The enhanced greenhouse effect

before 200 years ago have to be explained by natural causes. However, since then, we know that **human activities have released a lot of greenhouses gases into the atmosphere, chiefly carbon dioxide, methane, nitrous oxide and chlorofluorocarbons**. And if we have increased the amount of greenhouse gases in the atmosphere, it is highly possible that we have been responsible for changing the climate. Remember we found out in Chapter 1 that, in the last 100 years, global temperatures have increased by nearly 1 °C. Climatologists also predict that by 2100, temperatures could rise by another 1–6 °C if greenhouse gases are not reduced. So what have we been doing to cause the enhanced greenhouse effect?

## Increased carbon dioxide ($CO_2$)

The **burning of fossil fuels such as coal** has resulted in huge amounts of carbon dioxide being released into the atmosphere. Before the invention of electricity coal was very important as it was used to power factories and trains and to warm homes. Although widespread use of coal in the developed world is less common today, fossil fuels are still burned in power stations to produce electricity.

There are **more vehicles** on the world's roads today than there have ever been. Over one billion cars alone were recorded in 2011. Most vehicles run on petrol, which is a fossil fuel. When it is burned it releases $CO_2$ into the atmosphere.

**Deforestation** is an increasing problem across the world, particularly in the rainforests. Trees are known as 'the lungs of our planet' as they take in carbon dioxide and release oxygen. Destroying large areas of forest each year means there are fewer trees to take in carbon dioxide, which means more carbon dioxide remains in the atmosphere.

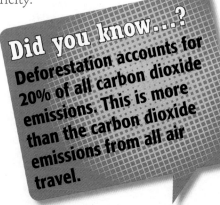

Did you know...?
Deforestation accounts for 20% of all carbon dioxide emissions. This is more than the carbon dioxide emissions from all air travel.

# CLIMATE CHANGE – HUMAN FACTORS

Figure 3.3
Burning coal increases $CO_2$ in the atmosphere

**Did you know...?** Each year 6 billion tonnes of carbon dioxide is released into the atmosphere.

## Increased methane ($CH_4$)

Each year 200 million tonnes of methane is released into the atmosphere. Most of the increase in methane comes from **padi fields**, which alone contribute about 10% of all greenhouse gas emissions. Padi fields are flooded fields that are used to grow rice. They are most common in eastern Asia where rice is the main food crop and used to feed millions of people. Rice absorbs carbon from the atmosphere; however, if the plant is not able to use the carbon it is dispersed into the soil where it converts to methane. It is then released into the atmosphere.

**Animal dung and belching cows** are major contributors to climate change. On average one cow releases about 100 kg of methane each year. Considering there are at least 1.5 billion cows on the planet, that is a lot of methane being released into the atmosphere. Much of the deforestation in the tropical rainforests has taken place in order to increase food production and much of the deforested land is being used for large-scale cattle ranches.

**Landfill sites** are huge areas of land that have been dug out so that domestic and industrial waste can be buried. As the waste begins to decompose it produces methane, which then goes into our atmosphere.

## Increased nitrous oxide ($N_2O$)

Nitrous oxide is 200–300 times more effective in trapping heat than carbon dioxide and it has one of the longest atmosphere lifetimes of all the greenhouse gasses, lasting for up to 120 years. Since the Industrial Revolution, the level of nitrous oxide in the atmosphere has increased by 16%.

**Did you know...?** The United Nations believes that 13 million hectares of forest is cut down each year. That is equivalent to 26 million football pitches!

Nitrous oxide is released when people add nitrogen to the soil by using **fertilisers**. **Soils release over half of the nitrous oxide in the atmosphere: two to four million tonnes per year.**

**Vehicles** are also a major contributor to increased levels of nitrous oxide, as it is released when petrol and diesel are burned.

### Increased chlorofluorocarbons (CFCs)

Chlorofluorocarbons (CFCs) are chemical compounds that were developed in the 1930s for use in **refrigeration** and **aerosols**. They are also used for **air-conditioning systems** and **polystyrene packaging**. The use of CFCs has been linked to the reduction of the ozone layer but the compounds found in CFCs are also greenhouse gases, which can be more harmful than carbon dioxide.

## National 4

1. What is the enhanced greenhouse effect?
2. Until 200 years ago, what caused all the changes to the world's climate?
3. Look at Figure 3.2.
   (a) What percentage of all the greenhouse gas emissions is carbon dioxide?
   (b) What are the other main greenhouse gases?
4. What human activities increase carbon dioxide in the atmosphere?
5. Three human activities are increasing methane in the atmosphere. Choose one and describe it.
6. What causes most of the extra nitrous oxide that goes into our atmosphere?
7. Why is nitrous oxide a particularly bad greenhouse gas?
8. What are CFCs and where are they found?

## National 5

1. What is the difference between the natural greenhouse effect and the enhanced greenhouse effect?
2. Look at Figure 3.2 and describe the relative importance of the greenhouse gases produced by human activity.
3. Describe the human activities which have increased the amount of carbon dioxide in the atmosphere.
4. Which is responsible for more greenhouse gases – crop farming or animal farming? Explain your choice.
5. Describe CFCs and explain where they are found.

# CLIMATE CHANGE – HUMAN FACTORS

## Activities

### Activity A

Look carefully at the pictures below and for each state:

(a) what the activity is
(b) what gas/gases the activity is releasing into the atmosphere.

A

B

C

D

E

F

## Activities continued...

G

H

I

### Activity B

Draw a table similar to the one below. List all the main human activities responsible for the enhanced greenhouse effect. Next to each one, write down whether you think the activity is increasing or decreasing, and why. The first activity has been completed for you.

| Human activity causing extra greenhouse gases | Increasing or decreasing |
|---|---|
| Burning fossil fuels | |

Now complete the 'I can do' boxes for this chapter.

# Chapter 4

*This chapter looks at the positive and negative effects of climate change.*

# Climate change – its effects

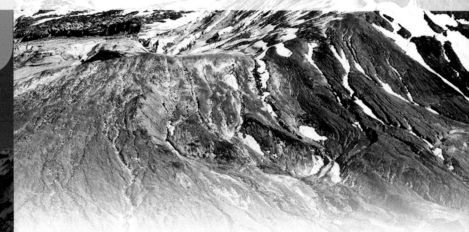

**By the end of this chapter, you should be able to:**

- ✓ give examples of some of the ways that changes in the climate can positively affect countries
- ✓ give examples of some of the ways that changes in the climate can negatively affect countries.

## Climate change and global warming

**Global warming** refers to the fact that the world has warmed by nearly 1 °C in the last 100 years. Most scientists believe it will rise by at least 2 °C in the next 100 years. Different regions are warming at different rates.

**Climate change** refers to the fact that it is not just the temperature which is changing; the amount of rainfall is also changing, leading to droughts and floods, and so is wind speed, leading to stronger storms. Different regions of the world are affected in different ways.

## Positive effects of climate change

Whenever climate change is mentioned, people think of its negative effects; however, there are some reasons why climate change can be thought of positively. Figure 4.1 shows some of these more positive effects. You can see that northerly latitudes benefit most. Many of these areas that benefit are in developed countries.

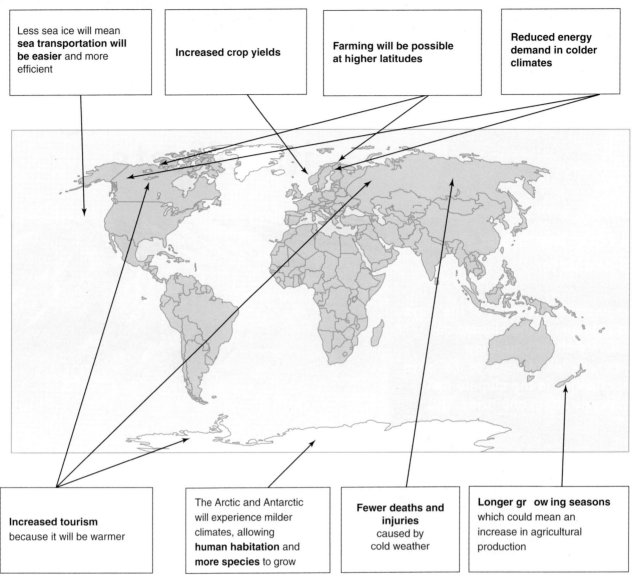

**Figure 4.1**
Positive effects of climate change

## Negative effects of climate change

Some of the negative effects of climate change are already being witnessed across the world, including floods, severe storms, droughts and food shortages. Tropical latitudes are particularly affected and most of these areas are in developing countries. Some of the negative effects of climate change are shown in Figure 4.2.

# CLIMATE CHANGE – ITS EFFECTS

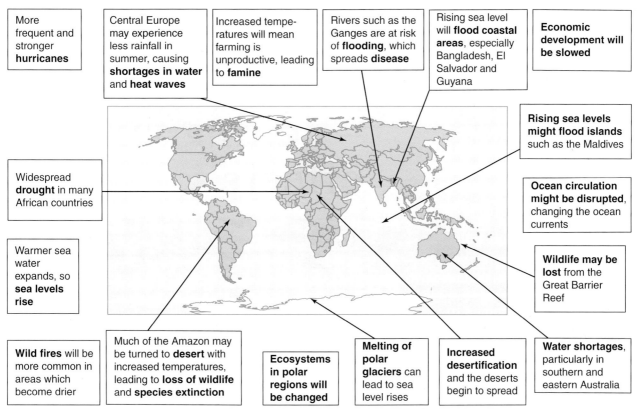

**Figure 4.2**
Negative effects of climate change

## Activity

Write a report on the global effects of climate change.

(a) Begin by briefly describing how the climate is changing and how it will change in the future.
(b) Then, describe and explain what you think are (i) the **three** biggest positive effects, and (ii) the **three** biggest negative effects.
(c) Try and expand on the points made in Figures 4.1 and 4.2. For example, if an area suffers from a severe drought, you can go on to say that crops and animals will die, there will not be enough water, people will have to move away, cities will grow in size, etc.
(d) Make sure you mention specific regions and countries and try and find out more details about them.
(e) Include diagrams and maps in your report.

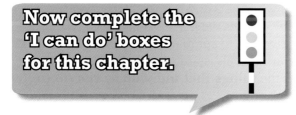

Now complete the 'I can do' boxes for this chapter.

# Chapter 5

This chapter looks at how climate change can be managed.

# Climate change – coping with its effects

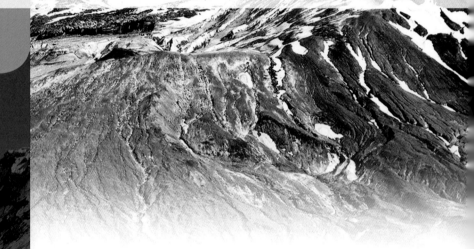

**By the end of this chapter, you should be able to:**

- ✓ give examples of strategies to reduce the effects of climate change at a local level
- ✓ describe some strategies that can be used on a national level to reduce the effects of climate change
- ✓ give examples of strategies to reduce the effects of climate change on an international level.

## Dealing with climate change

All scientists agree that global temperatures will continue to rise over the next 50 years at least. And this means there will be more floods, droughts, hurricanes, storms, etc. Nearly every country in the world has helped to cause climate change and nearly every country is affected by it. All countries must therefore take responsibility for reducing their harmful effects and every person should play his or her part in helping to reduce climate change.

## What can be done at a local level?

You may think that your activities and lifestyle have a tiny effect on climate change in comparison to large-scale activities such as cutting down forests and burning coal in power stations; in fact the choices we make play a major role in climate change. If we are to combat climate change, changes have to be made at all levels, starting with the individual. Here are some of the **things that individuals can do to reduce climate change**.

# CLIMATE CHANGE – COPING WITH ITS EFFECTS

1. **Reduce, Reuse, Recycle.** Recycling everyday items such as newspapers and milk containers as well as composting unused and wasted food reduces the amount that is sent to landfill sites and therefore reduces greenhouse gas emissions.
2. **Use less hot water.** It takes a lot of energy to heat water; reducing the settings on dishwashers and washing machines reduces the amount of energy needed.
3. **Insulate your home.** A lot of energy escapes through the roofs of homes that have not been well insulated, so people keep their central heating on for longer or use it more throughout the day. Central heating systems burn fossil fuels; adding more efficient insulation to your home reduces the need to use central heating and therefore reduces greenhouse gas emissions.
4. **Use energy-efficient light bulbs.** Energy-efficient light bulbs use 75% less energy than regular bulbs.
5. **Turn electrical equipment 'Off'.** It is very common for people to put the TV and other electrical equipment on standby rather than actually turning it off. By using standby the equipment is still using a great deal of energy.
6. **Turn lights off.** If you are not using a room in your home, turn the lights off.
7. **Leave the car at home.** Use public transport where possible rather than taking your car, or walk or cycle short distances.
8. **Cycle to school/work.**

If enough people decide they want to change their habits, companies and businesses will take notice. They will start producing and selling more energy-efficient products; company research departments as well as universities will start inventing new products to meet demand. The government will help to fund this research because it knows it will win votes. Small actions can therefore have a big effect.

## What can be done on a national level?

The **UK government** needs everyone to take action themselves, but it also knows it can and must take action as well. As a result, the **government has devised a set of climate change policies**, shown in Table 5.1.

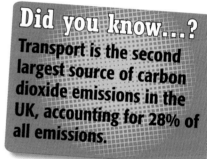

**Did you know...?** Transport is the second largest source of carbon dioxide emissions in the UK, accounting for 28% of all emissions.

| UK government policy | What does it mean? |
| --- | --- |
| The 2050 Challenge | The UK is committed to reducing greenhouse gas emissions by 80% by 2050. |
| Carbon budgets | This is a restriction on the total amount of greenhouse gases the UK can emit over a five-year period. |
| Green Deal | This makes it easier for householders and businesses to pay for some or all of the cost of energy-saving improvements to their properties over time. |
| CRC Energy Efficiency Scheme | This aims to improve energy efficiency and cut gas emissions in large organisations by requiring them to buy 'allowances' for every tonne of carbon they emit. |
| Encourage the use of ultra-low emission vehicles | The government provides grants to those who purchase electric, plug-in hybrid and hydrogen-powered cars and vans. It also provides funding to the Plugged-in Places programme. |
| Energy Targets | The UK has energy targets which it is committed to achieve. By 2020, 15% of our energy must come from renewable sources. There are other targets too, for instance on the emissions of carbon dioxide. |

*continued*

| Charge on single-use carrier bags | This could lead to a 90% reduction in carrier bags with 12 billion fewer plastic bags in circulation. |
|---|---|
| Carbon Emissions Reduction Target | Energy companies are obliged to give their customers better deals for being energy efficient. |
| Tax on 'gas guzzlers' | Cars that emit very low levels of carbon dioxide will pay no road tax; the more carbon dioxide cars emit, the greater the road tax levied on them. |
| Renewable Transport Fuel Obligation | Encourages the use of sustainable biofuels. |

**Table 5.1**
UK government climate change policies

## What can be done on an international level?

Climate change is global. Countries which cause global warming by burning large amounts of fossil fuels are not the only ones affected by it – the effects are global. International action to tackle climate change is therefore also needed. Two **examples of international strategies** are described below.

### United Nations Framework Convention on Climate Change (UNFCCC)

The UNFCCC is an international treaty set up in 1994 to tackle greenhouse gas emissions. Countries meet each year (in Conferences of the Parties) to decide plans. In 1997 in Kyoto, Japan, the countries agreed to cut their emissions by 5% by 2012. Some countries did not achieve this but, overall, emissions were reduced. In Paris in 2015 (COP21) 196 countries adopted the Paris Agreement. They agreed to keep global temperatures to below 2°C above pre-industrial levels and aim to keep them to 1.5°C above. In 2017, the USA decided to withdraw from this agreement.

### European Union legislation

The EU has committed to cutting its emissions by 20% by 2020. There are a number of EU initiatives to reduce greenhouse gas emissions, including the European Climate Change Programme (ECCP) and the EU Emissions Trading System. They agreed that by 2020 at least 20% of all the energy used in the EU would be from renewable energy sources. They also made targets to reduce carbon dioxide emissions from new cars and vans, and to support carbon capture and storage technologies.

**Did you know...?** The six largest emitters of carbon dioxide (USA, China, Europe, Russia, Japan and India) accounted for 70% of all energy-related carbon dioxide emissions in 2005.

## National 4

1. Choose three things that individuals can do to combat climate change and state why they are important.
2. Choose two of the UK's climate change policies that you think will be most successful and describe what they aim to do.
3. Describe the Kyoto Protocol.
4. Why do you think developing countries did not have targets under the Kyoto Protocol?

## National 4 continued...

5. Different countries have different targets to reach for reducing greenhouse gas emissions.
   (a) What is the UK's target for 2050?
   (b) What is the EU's target for 2020?
   (c) What was the UN's target for 2012?
6. Why do individuals have a responsibility to reduce the effects of climate change?
7. Which are likely to have the biggest effect on climate change – local actions, national actions or international actions? Give reasons for your answer.

## National 5

1. Choose three things that individuals can do to combat climate change and explain, in detail, why they are important.
2. Choose two of the UK's climate change policies that you think will be most successful and describe, in detail, what they aim to do.
3. Describe the purpose of the Kyoto Protocol and the Doha Amendment.
4. Why do you think developing countries do not have binding targets under the Kyoto Protocol (and the Doha Amendment)?
5. In your own words, describe in detail why it is the responsibility of each individual to reduce the effects of climate change.
6. Different countries have different targets to reach for reducing greenhouse gas emissions. What are the UK's national and international targets?
7. Which are likely to have the biggest effect on climate change – local, national or international actions? Give reasons for your answer.

## Activities

### Activity A

Design a leaflet to be posted to every household in the UK, showing how greenhouse gas emissions can be reduced. Your leaflet needs to be striking so people read it and don't just throw it away. It also needs to be powerful so people take action as a result.

### Activity B

1. Eight local actions are listed in this chapter. Write down which ones will cost you (and your family) money and which ones will save you money.
2. Ten UK climate change policies are listed in Table 5.1. Which policies will cost the government money and which policy will cost the most?

**Now complete the 'I can do' boxes for this chapter.**

# Chapter 6

## Climate change – case study of Bangladesh

This chapter looks at the effects of climate change on a developing country – Bangladesh.

**By the end of this chapter, you should be able to:**

- ✓ describe the human and physical geography of Bangladesh
- ✓ explain some effects of climate change on Bangladesh
- ✓ give reasons why Bangladesh finds it difficult to deal with the impact of climate change.

Figure 6.1
The country of Bangladesh

# Human geography of Bangladesh

| Population | 150 million |
|---|---|
| Population density | 1000 people per km² |
| GNI per capita | $1940 |
| Human Development Index position | 140th out of 177 countries |
| % of people employed in agriculture | 66% |
| Rural/urban population | 75% rural/25% urban |
| % people malnourished | 30% |
| Capital city | Dhaka |

Table 6.1
Human geography of Bangladesh

# Physical geography of Bangladesh

Bangladesh is a small country in South Asia. It has a land area of 147,000 km², of which 80% is flat floodplain. Three very large rivers flow through Bangladesh: the Ganges, the Brahmaputra and the Meghna, all of which join together to form the Ganges Delta before emptying into the Bay of Bengal.

Bangladesh is a very low-lying country with much of it at or near sea level. In fact, 90% of the whole country is less than 10 metres above sea level. Its only hills are in the southeast and northeast.

Bangladesh typically has three seasons: mild winters between October and March; hot, humid summers between March and June; and a warm, rainy monsoon season between June and October. It is one of the rainiest countries in the world and can suffer extreme river and sea floods. Bangladesh also suffers from hurricanes (cyclones) that affect the country between May and November, as well as occasional severe droughts.

Figure 6.2
Typical farming landscape in Bangladesh

# Effects of climate change on Bangladesh

Bangladesh is the most vulnerable country to climate change in the world. Over recent years the effects of global climate change have been very obvious here. Some of the effects that the country has witnessed are described below.

**Did you know...?** The sea level along the coast of Bangladesh is rising by about 3 mm each year.

## Rises in sea level

Sea levels around the world are rising every year because of polar ice melting (in particular the Greenland ice sheet) and warm water expanding. In 2000, the World Bank estimated that **Bangladesh would see sea level rises of 10 cm by 2020**. By the end of the twenty-first century it is expected that sea level rises will reach at least one metre in Bangladesh. The devastating effects of sea level rises on Bangladesh are shown in Figure 6.3.

## Increase in hurricanes

It is thought that hurricane (cyclone) activity has increased globally over the past century. Some scientists say that the number of hurricanes

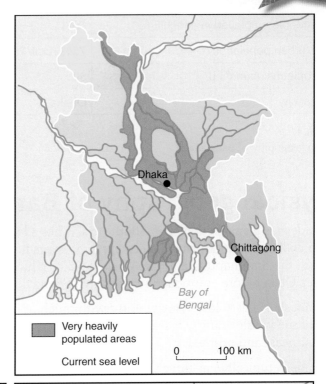

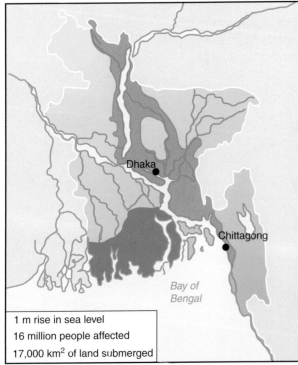

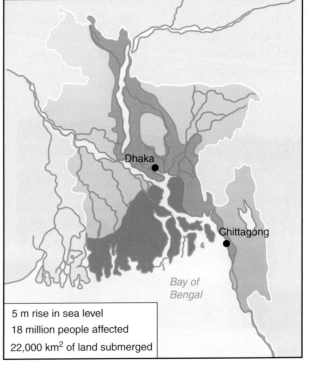

**Figure 6.3** Effects of sea level rise on Bangladesh

each year has doubled. This substantial increase in hurricanes is due to rising sea temperatures. Not only have hurricanes become more frequent but they are also increasing in intensity.

**Bangladesh now averages sixteen cyclones in every ten years.** Five of the ten worst cyclones in history have been in Bangladesh. In 2007, Cyclone Sidr killed 6000 people as it ripped through the coastal areas of Bangladesh. The following year Cyclone Aila killed a further 8000 people. Both Cyclone Sidr and Cyclone Aila are classed as 'super-cyclones' and to have two in two years is very unusual.

## Floods and flash floods

Bangladesh is a very small country but it has 250 rivers which flow over almost flat floodplains into the Ganges delta and the sea. It is therefore very vulnerable to floods. However, **the frequency and severity of the flooding is increasing**. In the nineteenth century, six major floods affected Bangladesh; in the twentieth century there were 18 major floods. Severe floods can cover over 60% of the country and remain there for several weeks. Bangladesh is the worst country in the world for flooding.

## Extreme temperatures and drought

Temperatures here can reach over 40 °C and, combined with the very high humidity in the rainy season, become almost unbearable. But now **temperatures are becoming even higher**, especially inland in northern and western parts. In addition, every few years Bangladesh suffers from drought which can quickly cause famine in the countryside.

Table 6.2 provides a summary of the problems and effects of climate change on Bangladesh.

| Problem | Effects |
| --- | --- |
| Sea level rises | Land is lost |
| | People lose their homes |
| | Farmland is destroyed |
| | Water shortages due to saltwater invasion |
| | Increased water-borne and water-related diseases |
| | Forced migration leads to 'climate change refugees' |
| | More people live in shanty towns |
| | Loss of mangrove forest |
| Increased hurricanes | Coastal devastation |
| | Houses, communications and businesses destroyed |
| | Loss of farmland |
| | Large-scale loss of life |
| | Cost of recovery prevents country developing |

Did you know…? In the last 130 years, Bangladesh has lost approximately 200 million people to 80 cyclones.

*continued*

| Problem | Effects |
|---|---|
| Flooding | People made homeless |
| | Loss of life |
| | Contamination of water leads to disease |
| | Food supplies affected as farmland and rice crops are destroyed |
| | Industries affected |
| | Communications can be badly damaged |
| Extreme temperatures and drought | Loss of life due to dehydration |
| | Famine |
| | Loss of livestock |
| | Land is unusable for farming |
| | Desertification |
| | Heatwaves cause fatalities |

**Table 6.2**
Impact of climate change on Bangladesh

# Managing climate change in Bangladesh

Although the effects of climate change can be seen the world over, they are most evident in the developing world, in particular Bangladesh. There are two reasons for this: first, the developing world is more dependent on the natural environment, and second, countries in the developing world do not have the means to protect themselves against the growing threat of climate change.

People living in Bangladesh are very heavily dependent on their natural environment through agriculture and fishing. With such a high percentage of people employed in these industries, the effects of climate change impact on the whole country. To try to combat some of the effects of climate change on these industries, Bangladesh has:

- developed and introduced salt-resistant strains of rice
- replaced rice farming with shrimp farming in areas where saltwater has invaded farmland
- given $150 million to 'climate proof' agriculture, for example new sources of freshwater, cyclone shelters and more research into better crop strains.

Bangladesh is one of the poorest countries in the world and it is too poor to introduce sophisticated technologies that countries in the developed world might use. In order to try to fight the effects of climate change Bangladesh desperately needs help from other developed countries, which it believes it is owed as compensation for the global warming caused by these countries.

# CLIMATE CHANGE – CASE STUDY OF BANGLADESH

## National 4

1. Look at Table 6.1. How do you know that Bangladesh is a developing country?
2. Give three problems of Bangladesh's physical landscape that make it vulnerable to the impacts of climate change.
3. Describe the effects of sea level rises on Bangladesh.
4. What is a 'climate change refugee'?
5. Give examples of the effects of cyclones on Bangladesh.
6. Which do you think has the worse effects on the people of Bangladesh – floods or drought? Give reasons for your choice.
7. The sea is gradually taking over farmland in Bangladesh. Describe one way in which the country is coping with this problem.
8. Explain why developing countries such as Bangladesh find it difficult to manage the effects of climate change.
9. Why does Bangladesh believe that other countries should help it manage climate change?

## National 5

1. What evidence is there to show that Bangladesh is a developing country?
2. The physical geography of Bangladesh is quite unique. In what ways does the physical landscape make Bangladesh vulnerable to the effects of climate change?
3. Choose the two most serious effects of climate change on Bangladesh and explain how they will affect the country.
4. People in Bangladesh insist that climate change is not in the future; it is happening to them *now*. What is their evidence for this?
5. Explain, in detail, why Bangladesh, like other developing countries, finds it difficult to manage the effects of climate change.
6. Describe the ways in which Bangladesh is coping with its freshwater becoming salty.
7. Do you agree that other countries should help Bangladesh 'as compensation'? Give reasons.

## Activities

Activities for this chapter appear at the end of Chapter 7.

Now complete the 'I can do' boxes for this chapter.

# Chapter 7

## Climate change – case study of Florida, USA

This chapter looks at the effects of climate change on a developed country – Florida, USA.

**By the end of this chapter, you should be able to:**

- ✓ describe the human and physical geography of Florida
- ✓ explain some of the effects of climate change on Florida
- ✓ give reasons why Florida finds it easier to deal with the impacts of climate change.

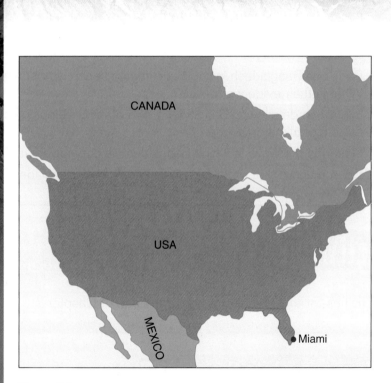

**Figure 7.1**
The location of Florida

# CLIMATE CHANGE – CASE STUDY OF FLORIDA, USA

## Human geography of Florida

| Population | 19 million |
|---|---|
| Population density | 353 people per km² |
| GNI per capita ($) | $48,890 (US average) |
| Human Development Index position | 3rd in the world (USA) |
| % of people employed in agriculture | 3% (US average) |
| Rural/urban population | 21% rural/79% urban (US average) |
| % people malnourished | N/A |
| State capital | Tallahassee |

**Table 7.1**
Human geography of Florida

## Physical geography of Florida

Florida is the most southeasterly state in the USA (Figure 7.1). It is situated on a peninsula between the Gulf of Mexico and the Atlantic Ocean and it borders the states of Georgia and Alabama to the north.

Florida is very low-lying, with approximately 50% of the land below 10 metres. In the south are the freshwater wetlands known as the Everglades (see Figure 7.2) which are home to many rare and endangered species such as the manatee, American crocodile and panther. Inland, in the north, Florida has a few hilly areas; however, the highest point is still only 105 metres above sea level.

**Figure 7.2**
Everglades National Park, Florida

Florida's climate is divided into two, with a warm temperate climate in the north, which has very warm, long summers and mild winters. The south of the state has a sub-tropical climate with hot and humid summers, a lot of rainfall and mild winters.

The state of Florida suffers natural disasters each year. From June to September Florida is the US state most at risk from hurricanes and it is also one of the most tornado-prone states in the USA.

## Effects of climate change on Florida

It is thought that climate change will have the following effects on Florida:

- Temperatures will increase by 2–4 °C over the next 80 years.
- Rainfall will become more intense.

- Droughts will increase, especially in summer.
- Sea levels will rise. **The sea level around Florida's coast is increasing at a rate of approximately 2 cm every ten years.** It is thought that by 2030, the sea level will have increased by approximately 12 cm and by as much as 1 metre by 2100. Eighty per cent of Florida's population lives in the coastal areas where the land is very flat. A 30 cm rise in sea level will move the shoreline inward by more than 300 metres. Figure 7.3 shows estimated sea level rises in Florida.
- **Hurricanes will become stronger.** Florida has seen an increase in intense hurricanes in recent years, such as Hurricane Katrina in 2005 and Superstorm Sandy in 2012. Scientists believe that the number of Category 4 and 5 hurricanes will increase by 80% by 2080. They also believe that the frequency of less intense hurricanes (Category 1–3) will decrease.

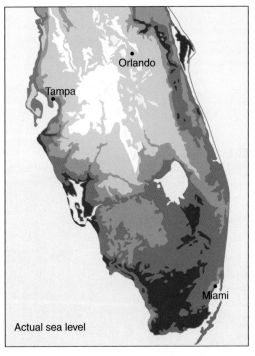
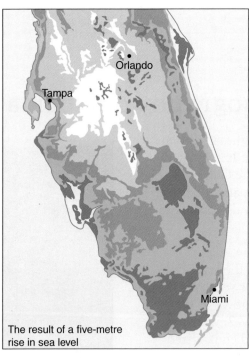

**Figure 7.3**
Effects of sea level rises in Florida

Table 7.2 provides a summary of the problems and effects of climate change on Florida.

| Problem | Effects |
| --- | --- |
| Sea level rise in Florida | Loss of land |
|  | Loss of mangroves |
|  | Loss of wildlife and ecosystems |
|  | Destruction of houses |
|  | Damage to water supply and drainage |
|  | Saltwater intrusion |
|  | Increased coastal erosion |
| Sea level rise in the Everglades | Saltwater intrusion will destroy the Everglades, causing economic losses from tourism of as much as $1 billion |
|  | Loss of many rare species of plants and animals |

*continued*

| Problem | Effects |
|---|---|
| Increased hurricanes and storm surges | Coastal devastation |
| | Destruction of homes and businesses |
| | Loss of life |
| | Damage to economy |
| | Fewer tourists |
| More intense rainfall | Widespread flooding |
| | Damage to property |
| | Damage to businesses/economy |
| | Loss of life |
| | Loss of crops |
| Drought in summer | Possible fatalities |
| | Loss of crops/farmland |
| | Water shortages |

**Table 7.2**
Impact of climate change on Florida

# How climate change is managed in Florida

The developed world is in a far better position than developing countries to deal with the effects of climate change. Although developed countries are significantly affected by changes in the world's climate, they have the money to protect themselves, which puts them in a better position. Florida is witnessing the effects of climate change now, and it has put several measures in place in order to combat these effects:

- Raising low-lying roads
- Adding protective beach dunes (see Figure 7.4)
- Obtaining underground water from further inland
- Increasing the use of public transport
- Encouraging the use of solar power and other green (environmentally friendly) energies
- Raising sea walls
- Storing more storm water to supplement drinking water supplies
- Planting more trees in urban areas
- Protecting farmland and open spaces from development.

**Figure 7.4**
Protecting sand dunes in Florida

## National 4

1. Look at Table 7.1. How do you know that Florida is a developed country?
2. In what ways does Florida's physical landscape make it vulnerable to the effects of climate change?
3. Describe the problems caused by sea level rises in Florida.
4. Explain how climate change will affect tourism in Florida.
5. Which areas of Florida will be affected by climate change? Give reasons for your answer.
6. Florida is trying to reduce the effects of climate change. Nine measures for doing so are mentioned in the text. Choose two of these measures and explain how each helps.

## National 5

1. What evidence is there to show that Florida is part of a developed country?
2. In what ways does the physical geography of Florida make it vulnerable to the effects of climate change?
3. Describe in detail the effects of rising sea levels on (a) the people and (b) the landscape of Florida.
4. Explain how climate change will affect the tourist industry in Florida.
5. Florida is trying to reduce the effects of climate change. Nine measures for doing so are mentioned in the text. Choose three of these measures and explain in detail how each helps.

## Activity

Using all the information in Chapters 6 and 7, design a large information poster comparing the effects of climate change on Bangladesh and Florida. Your poster should include information for each area on:

- (a) its location
- (b) its physical geography
- (c) its wealth
- (d) the effects of climate change
- (e) the ways it is managing the effects of climate change.

Try to design the poster so that it is easy to compare the two areas. Your poster should also include a range of presentation methods, such as maps, graphs, diagrams and pictures.

Now complete the 'I can do' boxes for this chapter.

# N4 Assessment question

Read the information and the assessment question below, and answer Tasks A and B.

## Question 1

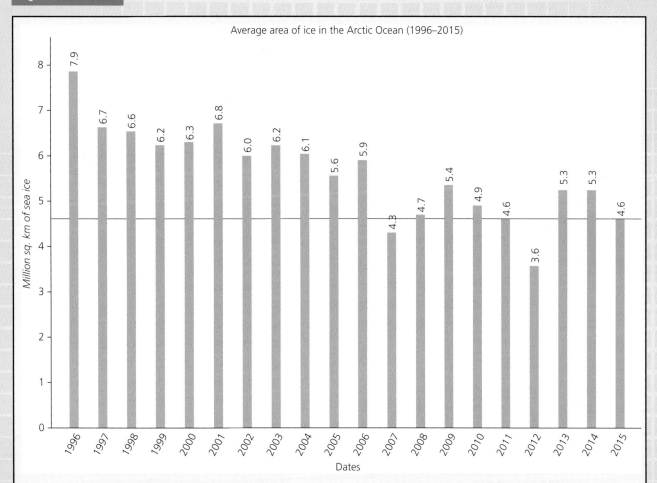

**Figure 7.5**
Diagram Q1 – Arctic sea ice changes (1996–2015)

(a) Look at Diagram Q1. Describe the changes in the area of Arctic sea ice between 1996 and 2015. (2)

(b) Experts believe the area of Arctic sea ice is changing because the world's climate is changing. Give reasons why the world's climate is changing. (3)

## Advice for Question 1(a)

You should have spotted three important points.

1 It is a *describe* question.
2 It has a diagram (Diagram Q1) which you must use.
3 It is worth **2** marks.

Let's take these points separately.
1 It is a *describe* question.
- This is one of two types of question you can be asked.
- For a *describe* question you must give facts. In this case you must say in what ways the area of Arctic sea ice has changed.

2 It has a diagram.
- Some *describe* questions do not have a diagram. For this type of *describe* question, you are given a diagram (the graph), which you must interpret.

3 It is worth **2** marks.
- There is 1 mark for each fact you give but you can earn 2 marks by giving a detailed fact.
- When describing a graph, you can earn more marks by giving figures. For example, *Between 1996 and 2015 the area of Arctic sea ice went down (1) from 7.9 million sq. km to 4.6 million sq. km. (1)*

**TASK A**: Read the advice for Question 1(a) and then answer Question 1(a).

## Advice for Question 1(b)

You should have spotted three important points.

1 It is a *give reasons* question.
2 It has no diagram.
3 It is worth **3** marks.

Let's take these points separately.
1 It is a *give reasons* question.
- This is one of two types of question you can be asked.
- You need to make a number of points that make the situation clear – in this case, you need to make it clear why our climate is changing.

2 It has no diagram.
- You have to answer this question from your own knowledge of climate change. You should have studied both physical and human causes of climate change and you can write about both for this question. Physical causes include volcanoes and sunspots; human causes include fossil fuels and deforestation.

3 It is worth **3** marks.
- There is 1 mark for each valid reason you give but you can earn 2 marks by giving an expanded reason. For example, *Our climate is becoming warmer because we burn a lot of fossil fuels such as coal and oil. (1) We burn a lot of oil when we drive cars and we burn a lot of coal to make electricity. (1)*

**TASK B**: Read the advice above and then answer Question 1(b).

# N5 Examination questions

Read the information and the two SQA-style questions below, and answer Tasks C–J.

## Question 2

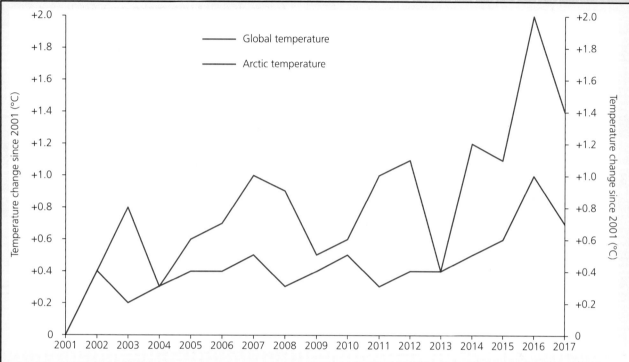

**Figure 7.6**
Diagram Q2 – Global and Arctic temperature changes (2001–2017)

**(a)** Study Diagram Q2. It shows temperature changes since 2001 for the world (global) and for the Arctic region. Describe, in detail, the differences in Global and Arctic temperatures between 2001 and 2017. (4)

**(b)** Describe, in detail, the effects of recent temperature changes on the Arctic region or any other area you have studied. (6)

### Advice for Question 2(a)

You should have spotted four important points.

1 It is a *describe* question.
2 It asks you to *describe differences*.
3 It has a diagram.
4 It is worth 4 marks.

### Advice for Question 2(b)

You should have spotted four important points.

1 It is another *describe* question, so you must state some factual points.
2 It has no diagram. For this type of *describe* question, you need to use your knowledge of the effects of recent temperature changes.

## Advice for Question 2(a)

Let's take these points separately.

1 It is a *describe* question.
- For a *describe* question you must make a number of factual points. In this case, you need to state a few similarities and differences in the temperature changes.

2 It asks you to *describe differences*.
- You score no marks for describing each graph separately. You only score marks by stating differences between them. For instance, *Arctic temperatures have risen faster than global temperatures.* (1)

3 It has a diagram.
- Some *describe* questions have no diagram. Other *describe* questions, like this one, have a diagram which you must interpret. It is usually a map or graph.

4 It is worth **4** marks.
- You score 1 mark for each difference you describe or 2 marks if it is a more detailed point. In interpreting a graph, you should always aim to quote figures. For instance, *Arctic temperatures rose by 1.4°C between 2001 and 2017 while global temperatures rose by half of that, 0.7°C.* (1)

**TASK C**: Read the advice for Question 2(a) and then answer Question 2(a).

## Advice for Question 2(b)

3 It asks you to *choose a region*. You should name your chosen area at the start of the question.

4 It is worth **6** marks. You score 1 mark for each point you make or 2 marks for a developed point. For a 6-mark question, you should try to make developed points. For example, *In Florida, the temperatures are rising which causes the sea level to rise. The rising sea level will flood over land and houses and kill wildlife and plants.* (2)

**TASK D**: Read the advice and then answer Question 2(b).

# Question 3

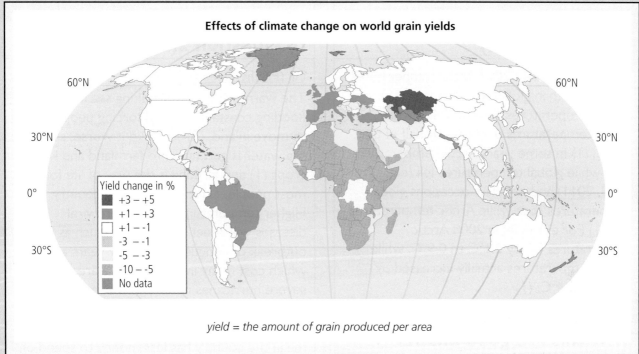

**Figure 7.7**
Diagram Q3 – Climate change and world grain yields

(a) Describe the effects of climate change on world grain yields, as shown in Diagram Q3. (4)

(b) Referring to areas you have studied, describe different ways in which climate change can be managed. (6)

## How good is this answer to Question 3(a)?

Some areas of the world will see a small increase in yields of grain, but far more areas will see a decrease in yields. This could lead to food shortages and famine which could cost many lives. However, some areas appear not to be affected at all.

**TASK E**: Read the answer above and give it a mark. Explain the number of marks you have given.

## How good is this answer to Question 3(b)?

The regions I have studied are dealing with climate change by reducing their carbon emissions. They are using fewer fossil fuels and more renewable energy sources. They are banning harmful substances from going into the atmosphere. Electric cars are being developed and people cycle more. They also recycle things more. This all helps.

**TASK F**: Read the answer above and give it a mark. Explain the number of marks you have given.

## Question 2 on page 39

### This is a good answer to Question 2(a)

Between 2001 and 2017 Arctic temperatures rose faster than global temperatures. (1) Arctic temperatures rose by 1.4°C while global temperatures only rose by half of that, 0.7°C. (1) In some years Arctic temperatures rose while global temperatures fell (e.g. 2003, 2011) and on a few occasions global temperatures rose while Arctic temperatures fell. (1) For example, in 2008 Arctic temperatures fell by 0.1°C to 0.9°C, while global temperatures actually increased by 0.1°C to 0.4°C. (1)

### Why this is a good answer

It makes a few factual points, enough to earn full marks. The points it makes all relate to differences between the two graphs. It gives precise information, mentioning specific years and temperatures. This shows the candidate can interpret the graph thoroughly.

**TASK G**: To make the answer even better, improve the third sentence by adding temperatures as well: *In some years Arctic temperatures rose while global temperatures fell (e.g. 2003, 2011) and on a few occasions global temperatures rose while Arctic temperatures fell...*

### This is a good answer to Question 2(b)

<u>Fiji (Pacific Ocean)</u>
The warmer climate is making sea level rise, flooding coastal areas and forcing people to move inland. (2) To make it worse, salty sea water is getting on to farmland and killing crops (1) and farming is the way of life for most people. (1) It is just a poor country. Higher temperatures mean more viral diseases will affect people here, such as the outbreak of dengue fever a few years ago which cost many lives. (1) There are more severe hurricanes now. The warmer climate is killing the lovely coral reefs around Fiji. (1) This will reduce the number of tourists and mean the country has less money to spend on improving people's standard of living. (2)

### Why this is a good answer

The candidate starts by naming the area they are describing. He/she makes a lot of factual points, which go into detail on the different effects. The question asks how temperature changes are affecting the chosen area and the candidate has made it clear that all the effects on Fiji mentioned are due to temperature rises.

**TASK H**: Two sentences in the answer do not earn marks. Rewrite these two sentences and improve on the points they make to earn marks.

# Question 3 on page 41

### Advice for Question 3(a)

You should have spotted three important points.

1 It is a *describe* question, so you must make some factual points.
2 It has a diagram – Diagram Q3. You must interpret the world map to score marks.
3 It is worth **4** marks, so you must make 4 valid points.

### Advice for Question 3(b)

You should have spotted four important points.

1 It is a *describe* question, so you must make some factual points on how climate change can be managed.
2 It has no diagram – you are expected to use your own knowledge of climate change.
3 It tells you to *refer to regions you have studied*, so you must name specific areas, such as Florida, Bangladesh or the UK.
4 It is worth **6** marks, so you must make 6 valid points or 3 developed points.

## How good was the answer on p.41?

**(Mark: 1 out of 4)**

The candidate would score an overall 1 mark for stating the general pattern shown by the map. The second sentence earns no marks because it is not strictly answering the question. To earn more marks, the candidate would have to interpret this map more thoroughly. When interpreting a map such as this, consider mentioning locations (e.g. *Africa will see the biggest decrease in yields and the areas which will see an increase in grain yields are mostly above 30° from the equator*). Also mention numbers to show that you can interpret the key (e.g. *Europe will have an increase in grain yields of 1–3%*).

**TASK I:** Read the comments above and then write an improved answer to Question 3(a).

## How good was the answer on p.41?

**(Mark: 1 out of 6)**

This candidate knows some ways of managing climate change but has answered the question badly. Not every country is tackling the problem in the way the candidate describes. The examiners want to know what specific countries are doing and how these measures help. Not only does this candidate not mention specific areas of the world, but they do not say how recycling or electric cars or renewable energy will help to manage climate change. For example, it would be better to say, *The UK is using fewer fossil fuels, such as coal and oil, which will reduce the amount of carbon in the atmosphere.* (1) As a separate point, you could mention: *Carbon dioxide is a greenhouse gas which keeps the Earth warmer.* (1) Despite having some knowledge of this topic, it is unlikely that the candidate would have passed this question.

**TASK J:** Read the comments above and then write an improved answer to Question 3(b).

# Chapter 8

## Structure of the Earth

This chapter looks at the structure of the Earth.

By the end of this chapter, you should be able to:

- ✓ explain the meaning of the term *natural (environmental) hazard*
- ✓ describe the three main layers of the Earth
- ✓ describe the two types of the Earth's crust.

**Did you know...?** Almost half the people in the world have lived through a natural disaster in the last ten years.

## Natural or environmental hazards

**Natural hazards, also known as environmental hazards, are sudden events in nature that cause people problems.** The problems may be slight (e.g. snow blocking roads) or severe (e.g. forest fires destroying property) or catastrophic (e.g. volcanic eruptions, earthquakes, drought, floods and tropical storms, which may kill hundreds of people).

The worst natural hazards are called natural disasters and are thought to kill on average 130,000 people every year, 97% of whom live in developing countries. It is estimated that these disasters cause damage totalling £60 billion a year. This topic looks at the most serious environmental hazards.

## The Earth's structure

The Earth is made up of three layers: the core, the mantle and the crust.

# STRUCTURE OF THE EARTH

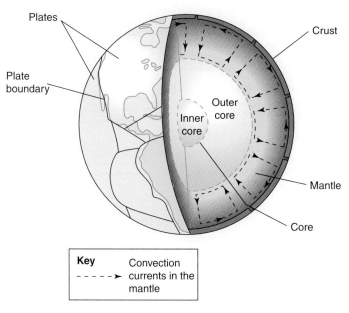

**Figure 8.1**
Structure of the Earth

**At the centre of the Earth is the core**, which is made up mostly of nickel and iron, where the temperature is thought to be between 4000 and 6000 °C. Very little is known about the core of the Earth because it is simply too hot.

**The layer that surrounds the core is called the mantle.** This is the thickest layer. The mantle is made up of molten rock (magma). Temperatures in the mantle can reach 3000 °C. Because the rocks in the mantle are molten or semi-molten they can move. They move because they are so hot. As the rocks heat up, they rise from the core to the crust, then cool down, spread out and sink back to the core (see Figure 8.1).

**The crust is the hard outside layer of the Earth.** It is the thinnest part of the Earth. If the Earth is compared to an apple, the crust would be as thin as the skin of the apple. The crust 'floats' on top of the mantle.

There are two types of crust: continental crust and oceanic crust. Continental crust is the oldest type of crust and is about 4 billion years old. The continental crust is the layer of rock that forms the continents. It is much thicker, lighter and older than that found under the oceans.

The oceanic crust is thinner, heavier and much younger; its oldest rocks are about 230 million years old, many are only a few million years old and some are only a few years old.

**The crust is not one single piece but is broken into many different pieces called tectonic plates.** These plates are constantly moving because they float on the molten rock in the mantle, and the molten rock is moving.

## Activities

### Activity A

Draw the structure of the Earth in your notebook. Colour the layers as follows:

- Core = Red
- Mantle = Orange
- Crust = Green

### Activity B

Read through the statements below about the structure of the Earth. Some of these statements are true and some are false. Identify what is wrong with each false statement and re-write it correctly.

1. The Earth is made up of three layers.
2. At the centre of the Earth is the crust. This can be split into two parts, the inner crust and the outer crust.
3. Temperatures at the centre of the Earth can be as high as 10,000°C.
4. The thickest layer is called the mantle.
5. The rocks in the mantle are constantly moving and are called convection currents.
6. If you compare the structure of the Earth to an apple, the crust would be the juicy middle bit.
7. There are two types of crust: continental crust and oceanic crust.
8. Continental crust is much thinner and younger than oceanic crust and is found under the oceans.
9. Oceanic crust is found under the oceans and all of it is about 230 million years old.
10. Continental crust is thicker than the oceanic crust, so it is heavier.

**Now complete the 'I can do' boxes for this chapter.**

# Chapter 9

## Crustal plates and plate boundaries

This chapter looks at crustal (tectonic) plates.

By the end of this chapter, you should be able to:

✓ give a definition of a crustal plate
✓ describe the four types of plate boundary
✓ describe the activities which take place at each plate boundary.

## Plate boundaries

Figure 9.1 shows that **the crust of the Earth is split into separate blocks called crustal plates**. The area where two plates meet is called a plate boundary. There are four types of plate boundary and different activities take place at each one.

## Destructive plate boundaries

Destructive plate boundaries (Figure 9.2) are found where an oceanic plate and a continental plate are moving together. One plate is made of oceanic crust and the other is made of continental crust. Because oceanic crust is heavier it is pushed downwards into the mantle where it melts, immediately forcing magma through the cracks to the Earth's surface and causing a volcanic eruption. Meanwhile, the surface rocks crumple together to form fold mountains. The rocks also crack as they are squeezed up, which triggers earthquakes.

This is taking place in Japan, in Indonesia and along the west coast of South America.

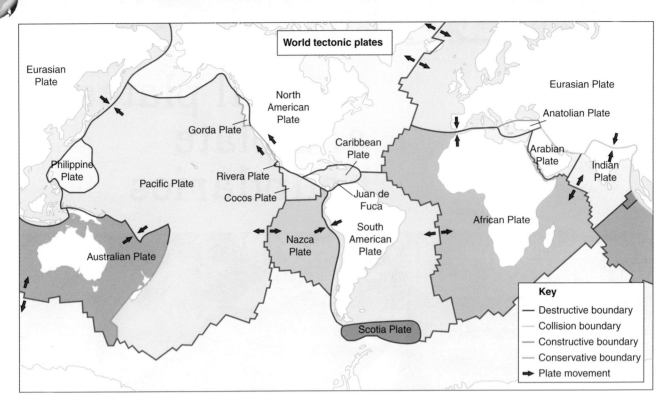

**Figure 9.1**
Crustal plates and plate boundaries

## Constructive plate boundaries

When two plates move apart, they form a constructive plate boundary (Figure 9.3). The crust cracks and splits, which allows magma from the mantle to reach the surface and cause a volcanic eruption. When the magma cools, it solidifies and forms new crust. Earthquakes also occur as the rocks split and move.

This is taking place in the middle of the Atlantic Ocean and the Southern Ocean near Antarctica.

## Conservative plate boundaries

When two plates slide past each other they form a conservative plate boundary (Figure 9.4). The sliding movement is not a smooth, continuous process; for most of the time the two plates are locked together. But, as pressure builds up, the plates suddenly jerk past each other causing an earthquake.

This is taking place in California, Turkey and New Zealand.

## Collision plate boundaries

Finally, when two continental plates move together, they form a collision plate boundary (Figure 9.5). The crust is too light to sink; instead it is forced upwards as the two plates come together. Over time this forms mountain ranges such as the Himalayas and the Alps. Earthquakes also occur as the rocks stretch and crack.

This is taking place in the Himalayas in Asia and the Alps and Atlas Mountains.

# CRUSTAL PLATES AND PLATE BOUNDARIES

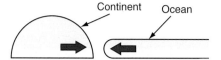

**Figure 9.2**
Destructive plate boundary

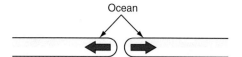

**Figure 9.3**
Constructive plate boundary

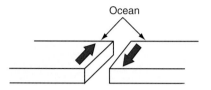

**Figure 9.4**
Conservative plate boundary

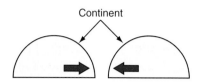

**Figure 9.5**
Collision plate boundary

## National 4

1. What is:
   (a) a crustal plate
   (b) a plate boundary?
2. What is the difference between a destructive plate boundary and a collision plate boundary?
3. Explain how the molten rock is able to reach the surface at a constructive plate boundary.
4. Draw a diagram to show a destructive plate boundary and label it to explain why earthquakes and volcanoes occur there.
5. Name the different plate boundaries where:
   (a) earthquakes occur
   (b) volcanoes occur.
6. Copy and complete Table 9.1 below. You may need to use an atlas to help you.

| Type of plate boundary | Description | Activity | Locations | Example |
|---|---|---|---|---|
| Collision | | | | Himalayas |
| | | Earthquakes<br><br>Volcanoes<br><br>Fold mountains | Nazca and South America | Andes |
| | Two plates move away from each other | Volcanoes<br><br>Earthquakes | | |
| | Two plates slide past each other | | | San Andreas Fault, USA |

Table 9.1

GLOBAL ISSUES

# National 5

1. Describe what is meant by a crustal plate and a plate boundary.
2. In your own words, summarise the movement of plates at each of the different plate boundaries.
3. Draw a diagram to show a destructive plate boundary and label it to explain why earthquakes and volcanoes occur there.
4. Copy and complete Table 9.1. You might need to use an atlas to help you.
5. Explain why fold mountains form at two types of plate boundary.
6. Explain why volcanoes are found at two types of plate boundary.

## Activities

### Activity A

Using an atlas, Figure 9.1 and the information in this chapter, find out near which type of plate boundary the cities/islands listed below are found. Choose from: destructive, constructive, conservative or collision.

| Reykjavik | Ankara | Jakarta | Tokyo |
| --- | --- | --- | --- |
| Lhasa | Lima | San Francisco | Christchurch |
| Algiers | St. Helena (island) | Kerguelen (island) | |

### Activity B

You will need a blank map of the world to complete this activity.

Scientists believe that all the continents were once joined together in a huge super-continent which they have called Pangaea. There is plenty of evidence for this.

(a) Complete the map using the information shown in the boxes below. You will need to draw a key underneath your map to show the information clearly.
(b) Imagine you are at a meeting where you must convince the audience that the continents of the world were once all joined together. Write your speech describing the evidence that our continents used to be joined together. Make the evidence as convincing as possible. At the end of the speech explain how it is possible for enormous continents to move across the world.

#### Matching rocks

There are igneous rocks in eastern India which are exactly the same as the igneous rocks in Western Australia. There are rocks containing iron in Western Australia exactly the same as the iron in rocks in South Africa.

Colour the rocks *red* along the:

- east cost of India
- west coast of Australia
- east coast of South Africa.

# CRUSTAL PLATES AND PLATE BOUNDARIES

## Activities continued...

### Lystrosaurus

*Lystrosaurus* was a dinosaur that lived about 200 million years ago. It died out long before humans appeared on Earth. It could not swim or fly, however fossils of *Lystrosaurus* have been found in:

- South Africa
- Antarctica
- India.

Colour each of these areas in *green*.

### The continents' shapes

The east coast of South America looks as if it fits into the west coast of Africa.

Colour each of these coastlines in *blue*.

### Matching mountains

The Appalachian Mountains run along the east coast of North America. The Scottish Highlands are exactly the same age and rock type as the Appalachians. So are the Norwegian Mountains. So are the mountains in Greenland.

Colour each of these mountain ranges in *brown*.

Now complete the 'I can do' boxes for this chapter.

# Chapter 10

*This chapter looks at volcanoes.*

# Volcanoes

**By the end of this chapter, you should be able to:**

- ✓ describe the location of volcanoes around the world
- ✓ explain the formation of volcanoes at plate boundaries
- ✓ describe the features of a volcano.

## Volcanoes as natural hazards

Volcanoes cause people many problems. Volcanic ash can cover houses and streets, lava can pour out over farmland and people may be forced to leave their homes when a volcano erupts. At their worst, volcanoes are killers.

The box below describes some of the worst volcanic eruptions in human history.

**Mt. Pelee (1902):** 28,000 people killed by a ball of lava that hurtled down the side of the volcano at 300 km/h.

**Vesuvius (AD79):** 2000 people suffocated by a massive downfall of hot volcanic ash that buried the town of Pompeii to a depth of 3 m in a very short time.

**Krakatoa (1883):** 36,000 people killed by tsunami up to 35 m high.

**Nevada del Ruiz (1985):** 20,000 people buried by a 40 m high mudflow (ash mixed with snow melt) sweeping down the volcano at 50 km/h, which then turned solid and trapped them.

# VOLCANOES

## Location of volcanoes

Figure 10.1 shows the location and distribution of active volcanoes in the world. **Active volcanoes are those that are likely to erupt**, for example Mt. Etna. Extinct volcanoes are those that will never erupt again, for example Edinburgh's volcano. There are also dormant volcanoes that have not erupted for at least 100 years, but may erupt again.

Active volcanoes are concentrated in just a few areas of the world. **Most are found near crustal plate boundaries.** In particular, they are located around the edge of the Pacific Ocean (e.g. Mt. St. Helens, Fujiyama), in the middle of the Atlantic Ocean (e.g. Surtsey) and through the Mediterranean Sea (e.g. Vesuvius, Etna).

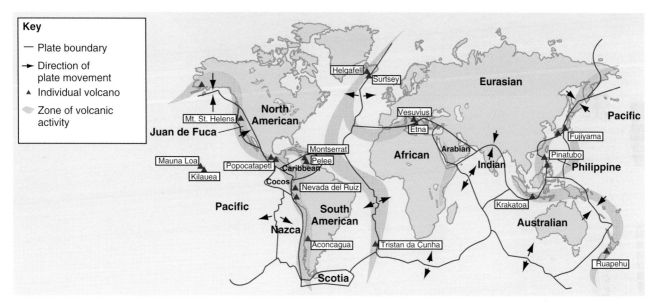

**Figure 10.1**
Distribution of plate boundaries and volcanoes

## Volcanoes at constructive plate boundaries

Constructive plate boundaries are found where two plates are moving apart (see Figure 10.2). As two plates move away from each other, liquid rock from the mantle rises through the cracks. When the magma reaches the surface an eruption occurs. On the surface, the lava begins to cool and becomes solid rock. This fills the crack where the two plates moved apart. As the plates continue to move apart, more and more cracks appear and the process repeats itself. Every time an eruption occurs, a layer of lava and ash is laid down over the land.

## Volcanoes at destructive plate boundaries

Destructive plate boundaries are found where two plates move together (Figure 10.3). As two plates move towards each other, the heavier oceanic plate is pushed downwards into the mantle, where it melts and the liquid rock (magma) makes its way through the cracks in the surface and erupts over the land. These eruptions can be very explosive.

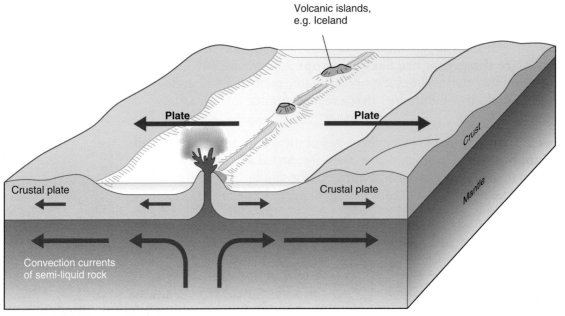

**Figure 10.2**
Volcanoes at constructive plate boundaries

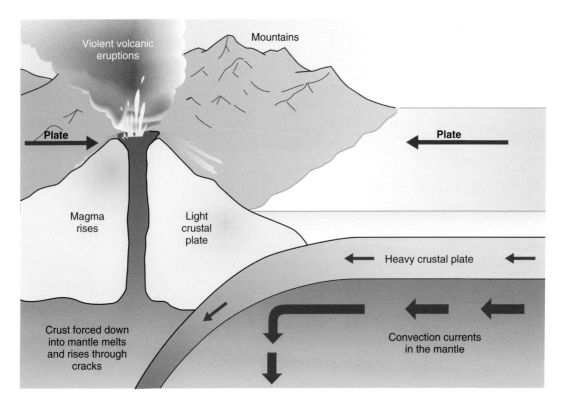

**Figure 10.3**
Volcanoes at destructive plate boundaries

# VOLCANOES

## Features of a volcano

**5** A violent eruption may also send lava into the air, which cools down, turns into solid rock and falls back to Earth as **volcanic bombs**.

**4** At the same time, the volcano explodes and rock is pulverised into tiny pieces called **volcanic ash**. When the ash falls back to the ground, it makes a separate layer on top of the lava.

**3** When the molten rock reaches the surface it pours out over the land and cools down, turning into solid rock.

**2** It then rises from the magma chamber through a crack to the surface. This crack is called a **vent**.

**1** At plate boundaries magma in the mantle begins to rise through cracks in crust. Here it collects in a **magma chamber**.

**7** Where the vent comes out of the top of the cone there is a large depression, called a **crater**.

**6** Each time the volcano erupts a layer of lava and ash is laid down so that, over time, a **volcanic cone** builds up.

**9** Where this vent reaches the surface, a **secondary cone** forms on the sides of the volcano.

**8** Once a volcano has erupted many times, the top of the vent may become blocked with solid lava. When the volcano next erupts, the magama cannot come out of the top and, instead, finds a weak area in the sides and forms a **secondary vent**.

**Figure 10.4**
Features of a volcano

## National 4

1. What is the difference between an active and a dormant volcano?
2. Describe the location of active volcanoes in the world.
3. Describe how volcanoes form at constructive plate boundaries. You should draw a diagram in your answer.
4. Describe how volcanoes form at destructive plate boundaries.
5. Name the six features described in the list below:
   (a) wide vertical crack inside a volcano
   (b) pool of liquid rock deep in the crust
   (c) lava blown into the air and turning solid
   (d) the shape of a volcano
   (e) depression at the top of a volcano
   (f) tiny pieces of rock coming out of a volcano
6. Look at Figure 10.5 below. Redraw this diagram and label the features listed in Question 5.

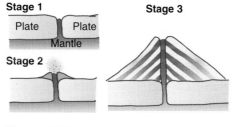

**Figure 10.5**

# National 5

1. Volcanoes are classed using names that describe how recently they erupted. What are these three names and what do they mean?
2. Using the information shown in Figure 10.1, describe, in detail, the location of active volcanoes in the world.
3. Using a diagram, explain why volcanoes are found at destructive plate boundaries.
4. Describe, in detail, how volcanoes form at constructive plate boundaries.
5. Name the six features described in the list below:
    (a) wide vertical crack inside a volcano
    (b) pool of liquid rock deep in the crust
    (c) lava blown into the air and turning solid
    (d) formed when the volcano's main vent is blocked
    (e) depression at the top of a volcano
    (f) made of layers of lava and ash
6. Look at Figure 10.5. Redraw this diagram, label the features shown and add a magma chamber and secondary vent.

# Activities

## Activity A

Using the information in the box describing notorious volcanic eruptions (on page 52), draw a bar graph showing the number of people that were killed during each volcano's eruption.

## Activity B

The following sentences describe how a volcano forms at constructive plate boundaries. However, the information is shown in the wrong order. Rearrange the sentences into the correct order.

- On the surface, the lava begins to cool and becomes solid rock.
- As the two plates move away from each other, liquid rock from the mantle rises through the cracks.
- As the plates continue to move apart, more and more cracks appear and the process repeats itself.
- Every time an eruption occurs, a layer of lava and ash is laid down.
- When the magma reaches the surface an eruption occurs.
- This fills the crack where the two plates moved apart.

## Activity C

The following sentences describe how volcanoes form at destructive plate boundaries. Some of the information in *each* sentence is incorrect and the sentences are not in the correct order. Correct the information in each sentence and then rearrange the sentences into the correct order.

# Activities continued...

- As it is pushed downwards, it solidifies.
- The solid rock (magma) then makes its way through the cracks in the surface and explodes.
- One of the continental plates is pushed downwards into the core.
- Two continental plates come together.

## Activity D

Make your own volcano!

You will need:

- Clay
- Paints
- Plastic bottle
- Warm water
- Red food colouring
- Washing-up liquid
- Baking soda
- Vinegar

### Instructions

1. Place the plastic bottle in the middle and mould your clay around it, to form a volcano shape.
2. Colour your clay with paint.
3. Pour the warm water into the plastic bottle and add a couple of drops of red food colouring.
4. Add six drops of washing up liquid to the bottle.
5. Add two tablespoons of baking soda.
6. Slowly pour the vinegar into the bottle and watch your volcano erupt!

Now complete the 'I can do' boxes for this chapter.

# Chapter 11

*This chapter looks at the eruption of Mt. St. Helens.*

# The eruption of Mt. St. Helens, 1980

**By the end of this chapter, you should be able to:**

- ✓ explain the reason for the eruption
- ✓ give examples of the effects of the eruption on the landscape
- ✓ describe the impact of the eruption on the people.

## Cause of the eruption

Mt. St. Helens is in the Rocky Mountains near the west coast of the USA, in the state of Washington. **It lies near to a destructive plate boundary** (see Figure 11.1), where the small Juan de Fuca Plate is moving southeast and the North American Plate is moving northwest.

**The small plate is being forced under the larger plate and into the mantle.** Here it melts, partly because of the heat and partly because of the immense friction as two plates grind together. **As it melts, molten rock rises into the crust.** Here it builds up in magma chambers until it is able to force its way through cracks in the crust to the surface (Figure 11.2). This has happened many times, for example at Mt. Lassen (also known as Lassen Peak) in 1914, Mt. Rainier in 1834 and, catastrophically, at Mt. St. Helens in 1980.

## The eruption

On 18 May 1980 Mt. St. Helens erupted for the first time in 123 years. It erupted with a power 500 times greater than any atomic bomb exploded during World War Two and was the

# THE ERUPTION OF MT. ST. HELENS, 1980

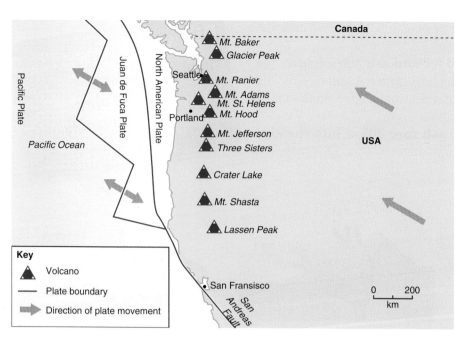

**Figure 11.1**
The location of Mt. St. Helens

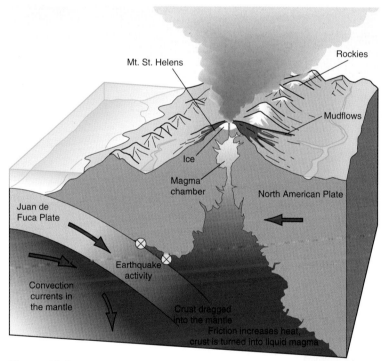

**Figure 11.2**
Cause of the eruption of Mt. St. Helens

most powerful eruption on Earth in the last 60 years. No lava poured out, but the event still had four devastating effects:

1. The eruption was triggered by the **biggest landslide in recorded history**, which sped down the north side of the mountain at 250 km/h.
2. There was a **tremendous blast** from the eruption, which could be heard 300 km away. The blast travelled at 500 km/h but this increased at times to twice that speed, overtaking the landslide. The blast contained rock,

ash and gases at temperatures of over 300 °C, known as a **pyroclastic flow**.
3. A **mudflow of rock, melted ice and ash** hurtled down the mountainside at 250 km/h. The heat from the eruption melted ice and snow on the mountain, releasing 200,000 million litres of water.
4. Around 400 million tonnes of **ash rose 20 km into the air**. Some rose so high, it never came down.

Did you know...?
Enough trees were flattened by the blast to build 300,000 family homes.

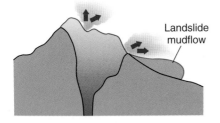
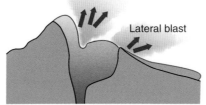
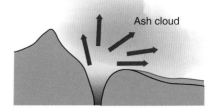

**Figure 11.3**
The eruption of Mt. St. Helens

## Impact on the landscape

- The landslide buried the North Fork Toutle River (see Figure 11.5) to a depth of 200 metres.
- The blast and pyroclastic flow killed every form of plant and animal life for a distance of 25 km north of the volcano. Even fully-grown fir trees were flattened, up to 30 km away. About 7000 animals died, including elk and bears.
- The mudflow choked rivers with sediment, killing all fish and water life and completely filling in Spirit Lake. About 12 million salmon died. The mud emptied itself into the sea at Portland, clogging up the harbour.
- The eruption of ash blew away the top of the mountain. In seconds it changed from a mountain 2950 metres high to one that was only 2560 metres high. At the top a crater 500 metres deep formed.

**Figure 11.4**
Trees flattened by the lava flow

## Impact on people

- The eruption on 18 May 1980 occurred on a Sunday, so no one was working in the forests that cover the slopes of Mt. St. Helens. Local people had been evacuated from their homes and tourists were prevented from getting close. In spite of all this, the eruption still killed 57 people and 198 had to be rescued. Damage ran into billions of dollars.
- **Mt. St. Helens had given clear warnings that it might erupt explosively.** From March onwards there had been minor earthquakes and small eruptions of ash and steam. These gradually became more severe.
- As a result, **the authorities were able to evacuate residents, tourists and forestry workers** from the surrounding area. They researched the area affected

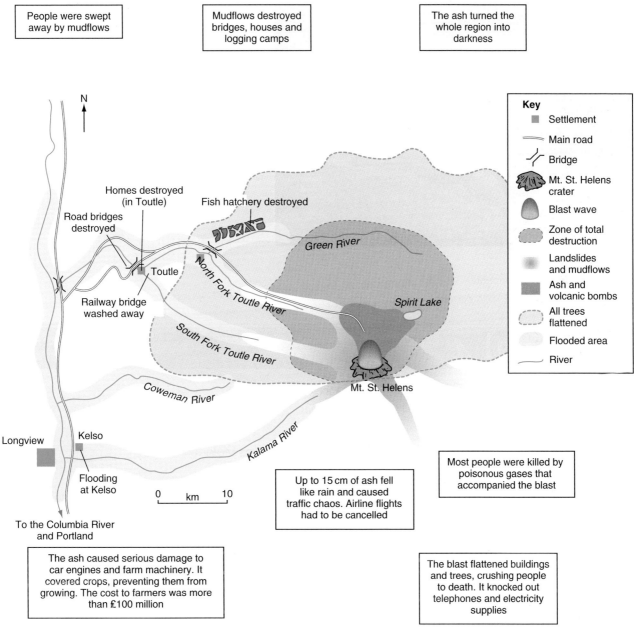

**Figure 11.5**
Effects of the eruption of Mt. St. Helens

by the previous eruption and made this an exclusion zone around Mt. St. Helens. **Emergency services were on hand**, including helicopters and aeroplanes.

- But the scientists could not give a precise date for the eruption. They tried measuring (a) the frequency of earthquakes on the mountain – the greater the frequency, the nearer the eruption, and (b) the size of the volcanic cone – the volcano bulged as magma built up in the vent. Even the day before the eruption, scientists were stating that the eruption might still be a few weeks away. **Nor did the experts predict that the blast from the eruption would be from the north side.** As a result, 90% of the people killed were outside the exclusion zone.

**Figure 11.6**
Vehicles trapped by the eruption

# National 4

1. Where is Mt. St. Helens?
2. When did the eruption of Mt. St. Helens take place?
3. Explain why the eruption took place.
4. Give three pieces of evidence to show that the eruption was violent.
5. Describe the effects on the landscape of the blast and the mudflow.
6. Which was the worst effect on the local people – the ash eruption, the mudflow or the blast? Give reasons for your answer.
7. Were the authorities able to predict that Mt. St. Helens was going to erupt? Give reasons for your answer.
8. How well did the authorities plan for the eruption?

# THE ERUPTION OF MT. ST. HELENS, 1980

## National 5

1. Describe the location of Mt. St. Helens and the date of its eruption.
2. Explain, in detail, why Mt. St. Helens erupted.
3. Copy the table below and fill in the details by writing the different effects of the eruption in the appropriate box.

|  | Effects of Mt. St. Helens eruption 1980 | |
|---|---|---|
|  | On the landscape | On the people |
| Ash eruption |  |  |
| Blast |  |  |
| Mudflow |  |  |

4. Describe the different methods of predicting the eruption of Mt. St. Helens.
5. Explain how effective these methods were.
6. Do you think the authorities were well-prepared for the eruption? Give reasons for your answer.

## Activities

### Activity A

Imagine the date is 17 May 1980 and you and your friends are arguing about whether to camp on the lower slopes of Mt. St. Helens. What would be the main arguments for and against camping?

### Activity B

After the argument you decide to camp on Mt. St. Helens and the next day the volcano erupts. Describe what you see, feel and hear.

Now complete the 'I can do' boxes for this chapter.

# Chapter 12

## Managing the eruption of Mt. St. Helens, 1980

*This chapter looks at how the people coped with the eruption of Mt. St. Helens.*

**By the end of this chapter, you should be able to:**

✓ give examples of help that was given before the eruption
✓ describe some of the short-term aid needed following the eruption
✓ describe what long-term aid is and why it was needed.

After any natural disaster, there is both official and voluntary aid to help the area recover. **Voluntary aid is provided by charities** and individuals; **official aid comes from governments**. Most aid usually comes from the home country but other countries may also send help.

After any natural disaster, there is a need for short-term aid and long-term aid. **Short-term aid is needed to rescue people** and then to provide emergency help for the survivors – food, clean water, medicine, tents and blankets. It is also needed to clear up the area and make it possible for people to survive.

**Long-term aid is needed to allow the area to return to normal.**

**Aid can be** in the form of **money**, but it can be **goods** (machinery, medicine, food) and it can also be **skilled people**.

The US federal government gave the most aid altogether, totalling $950 million.

# MANAGING THE ERUPTION OF MT. ST. HELENS, 1980

Figure 12.1
Vehicle buried by ash

Figure 12.2
Road block on Mt. St. Helens

- Part of the mountain was closed off before the eruption.
- About 2000 people were evacuated before and during the eruption.
- This saved many lives.
- However, people were still allowed to live and work near the mountain, and some refused to leave.
- Rescue centres were set up.
- 198 people were rescued.
- 57 people died.

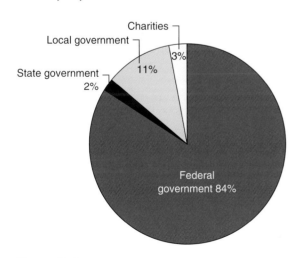

Figure 12.3
Aid given for the clean-up operation after the Mt. St. Helens eruption

Figure 12.4
Removing ash

- 1 million tonnes of ash were removed from roads and airports, at a cost of $1 million.
- 200,000 people were temporarily employed in the clean-up operation.
- But local people were unprepared for the ash fall or its effects, in particular on transportation and water treatment systems.
- Millions of trees were replanted.
- It will be 2050 before all the plants, trees and wildlife return to normal.

**Figure 12.5**
Replanted trees on Mt. St. Helens

Money was given by the US government to rebuild houses, repair roads and construct a new highway, costing $145 million.

The private National Science Foundation funded 74 research projects on the eruption over ten years, costing $5 million.

Federal and state money was given for new salmon hatcheries and farmers were also compensated because their crops had been ruined by the ash fall.

The Mt. St. Helens Visitor Center was built to attract tourists back to the area, together with more trails and information centres. This cost $50 million of federal money.

Local rivers were dredged to remove logs and levées were built up beside the Columbia River to reduce flooding.

Many charities and local people helped in the relief effort after the eruption.

## Activity

Write a report on the different types of aid given for the eruption of Mt. St. Helens in 1980, mentioning how effective each was.

Now complete the 'I can do' boxes for this chapter.

# Chapter 13

## Earthquakes

*This chapter looks at the location and features of earthquakes.*

**By the end of this chapter, you should be able to:**

- ✓ describe where earthquakes occur
- ✓ explain how an earthquake happens
- ✓ describe how earthquakes are measured.

## Earthquakes as natural hazards

Each year 20,000 people are killed by earthquakes, which makes them bigger killers than volcanoes. This is partly because earthquakes give no warning. It is also because many areas that suffer earthquakes are popular areas in which to live, such as California, and, in some cases, people do not know they are living in such a dangerous place. Earthquakes are also very common.

An earthquake happens somewhere in the world every two minutes. But most are very slight and they mainly occur under the sea. No one hears of them. Sometimes, however, there are severe earthquakes and, just occasionally, they take place under a large town. This is when earthquakes make headline news.

*31 May 1970*
**Earthquake in Peru Causes Landslide**
50,000 feared buried alive

*18 April 1906*
**Fires Destroy San Francisco**
caused by severe earthquake

*11 March 2010*
**Tsunami Alert**
as quake hits Taiwan

*12 January 2010*
**Chaos in Haiti**
200,000 feared dead in massive earthquake

## Location of earthquakes

The location of earthquakes (Figure 13.1) is very similar to that of volcanoes. They are concentrated in just a few parts of the world. **Nearly all take place near crustal plate boundaries.**

Earthquakes are particularly common around the edge of the Pacific Ocean (e.g. Japan, California) and through the Mediterranean Sea (e.g. Turkey, Italy). Most earthquakes occur under the sea (e.g. mid-Atlantic) because most plate boundaries are found there.

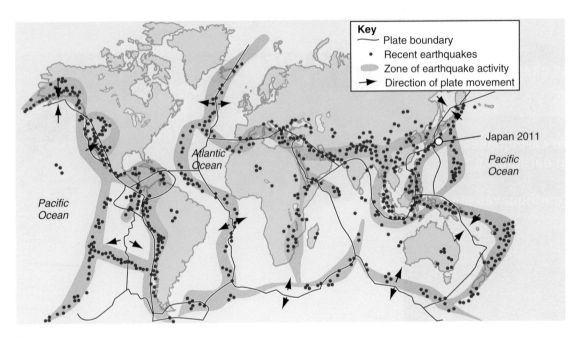

**Figure 13.1**
Distribution of earthquakes

## Features of an earthquake

**An earthquake occurs when rocks inside the crust move suddenly.** Where this happens is called the *focus* of the earthquake. **This sudden movement causes shock waves** to travel out in all directions. The place on the surface directly above the focus receives the worst shock waves. This is called the *epicentre* (Figure 13.2). **There are three types of shock waves:**

1. **P waves** (push or primary waves) make the rocks move up and down – they travel the fastest.
2. **S waves** (shake or secondary waves) make the rocks move from side to side – they travel at two-thirds the speed of P waves.
3. **L waves** (long waves) spread out in waves along the surface – they are the slowest but the most destructive.

The **shock waves are detected on seismographs or seismometers** (Figure 13.3). The magnitude of the earthquake is **measured on the Richter scale**. This is a logarithmic scale from 1–12. Earthquakes of scale 3 or under are minor and are not usually strong enough to be felt. Scale 6 or more are severe. No earthquake has yet registered a scale 10.

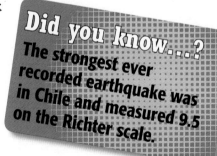

Did you know....?
The strongest ever recorded earthquake was in Chile and measured 9.5 on the Richter scale.

# EARTHQUAKES

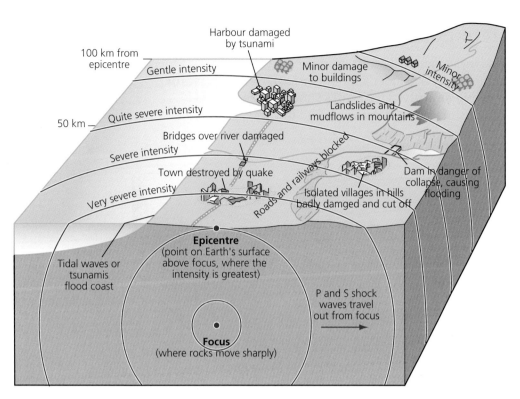

**Figure 13.2**
Features of an earthquake

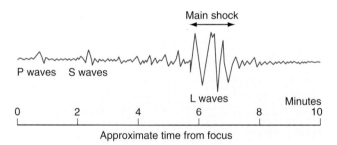

**Figure 13.3**
A seismograph

## The cause of earthquakes

**Earthquakes occur when rocks in the crust move suddenly.** This sets up shock waves that travel out in all directions. **This is most likely to happen at plate boundaries** where plates are trying to move in different directions. Earthquakes take place at three types of plate boundary: constructive, destructive and conservative plate boundaries (see Chapter 9).

## Constructive plate boundaries

At constructive plate boundaries (Figure 13.4) the crust is being forced in opposite directions. This puts the rocks under a lot of tension. Eventually, **some of the rocks crack and move sharply**. This causes shock waves, which travel through the crust to the surface, causing the ground to shake. This is the earthquake.

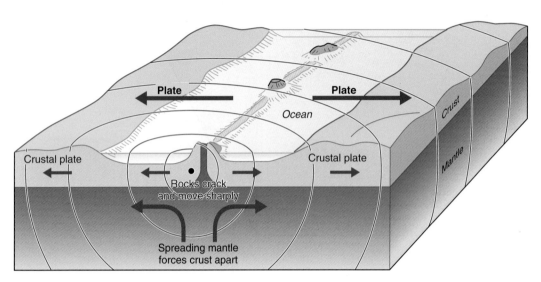

**Figure 13.4**
Earthquake activity at a constructive plate boundary (black dot represents the focus of the earthquake)

## Destructive plate boundaries

At destructive plate boundaries (Figure 13.5) **one crustal plate is being forced down below another**. But the friction between these huge chunks of crust is immense and stops the plates from moving. Eventually, however, the pressure continues to build up and **the crust jerks downwards into the mantle**. This sudden movement sends out shock waves that are felt as earthquakes on the surface. An example of where this occurs is Alaska.

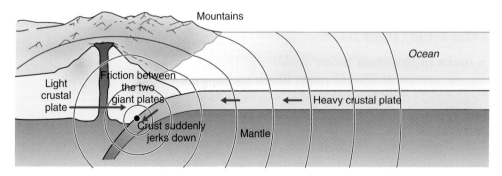

**Figure 13.5**
Earthquake activity at a destructive plate boundary (black dot represents the focus of the earthquake)

## Conservative (sliding) plate boundaries

In some areas of the world **where crustal plates meet, they just slide past one another**. No crust is destroyed or constructed and no volcanic activity takes place. But **the sliding movement is not smooth**. Because of the immense friction between the two slabs of crust, **the plates are locked together most of the time**. When the pressure has built up over a long period of time, it is great enough to overcome the friction and this is when **one plate suddenly jerks past the other**. This causes shock waves and an earthquake on the surface. An example of where this occurs is California.

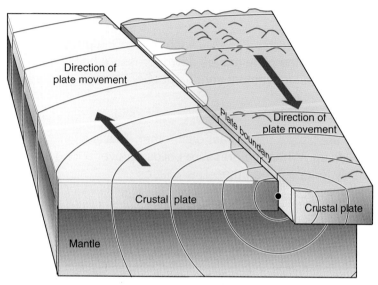

**Figure 13.6**
Earthquake activity at a conservative (sliding) plate boundary (black dot represents the focus of the earthquake)

# National 4

1. Why do earthquakes kill more people than do volcanoes?
2. Why do you think we only hear of the earthquakes that happen under large towns?
3. Describe the location of earthquakes.
4. The words in the list below and their definitions have been jumbled up. Match each word to its correct definition.

| | |
|---|---|
| Focus | The logarithmic scale showing the magnitude of the earthquake |
| Epicentre | The instrument used to measure the strength of the earthquake |
| Seismograph | The point where the rocks move suddenly inside the Earth's crust |
| Richter scale | The point on the Earth's surface above where the earthquake occurred |

# National 4 continued...

5. Name the three types of shock waves.
6. An earthquake is caused by shock waves at the Earth's surface. What causes the shock waves?
7. Copy and complete the paragraphs below describing how earthquakes occur at the three different plate boundaries. Use the word banks below to help you.
   At constructive plate boundaries, the plates are moving _____ from each other. As this happens, some of the rocks crack and move _____. This causes _____, which travel through the crust to the surface where they cause the ground to _____.
   
   *shake    sharply    away    shock waves*
   
   At destructive plate boundaries, the plates are moving _____ each other. One plate is forced down into the _____ by the other. As the plate moves down, pressure builds up and the plate _____ downwards. This sudden movement sends out shock waves that are felt as an _____ at the surface.
   
   *jerks    earthquake    towards    mantle*
   
   At conservative plate boundaries, the plates are _____ past each other. This is not a smooth process and _____ builds up. When the pressure builds up over a long period of time, one _____ will suddenly jerk past the other. This causes _____ and an earthquake on the surface.
   
   *pressure    plate    sliding    shock waves*

# National 5

1. A magnitude 8.2 earthquake occurred in the middle of the Pacific Ocean, while a magnitude 5.9 earthquake struck San Francisco. Which would be reported on the news in Britain – neither, one or both? Give reasons for your answer.
2. Describe the relationship between plates, plate boundaries and earthquakes.
3. In your own words, describe each of the following terms:
   focus, epicentre, seismograph, Richter scale.
4. Describe the three types of shock waves.
5. Explain, in detail, why earthquakes occur at constructive plate boundaries.
6. At destructive plate boundaries, the crust moves in a series of jerks. Explain why it moves in this way.
7. Draw an annotated diagram to explain how earthquakes occur at conservative plate boundaries.

EARTHQUAKES

## Activities

### Activity A

The strongest earthquakes ever recorded on the Richter scale are shown in this table. Draw a bar graph in your notebook to show the strength of these earthquakes.

|    | Where? | Magnitude | When? |
| --- | --- | --- | --- |
| 1 | Chile | 9.5 | 22 May 1960 |
| 2 | Prince William Sound, Alaska | 9.2 | 28 March 1964 |
| 3 | Off the west coast of Northern Sumatra | 9.1 | 26 December 2004 |
| 4 | Near the east coast of Honshu, Japan | 9.0 | 11 March 2011 |
| 5 | Kamchatka, Far East Russia | 9.0 | 4 November 1952 |
| 6 | Off the coast of Maule, Chile | 8.8 | 27 February 2010 |
| 7 | Off the coast of Ecuador | 8.8 | 31 January 1906 |
| 8 | Rat Islands, Alaska | 8.7 | 4 February 1965 |
| 9 | Northern Sumatra, Indonesia | 8.6 | 28 March 2005 |
| 10 | Assam, Tibet | 8.6 | 15 August 1950 |

### Activity B

Using Figure 13.1 and an atlas, try to work out which two plates caused each of the earthquakes shown in the table above.

**Now complete the 'I can do' boxes for this chapter.**

# Chapter 14

## The cause of the Japan earthquake, 2011

*This chapter looks at the 2011 Japan earthquake.*

**By the end of this chapter, you should be able to:**

- ✓ explain why Japan experiences earthquakes
- ✓ describe the cause of the 2011 earthquake
- ✓ give reasons why predicting earthquakes is difficult.

## The 2011 Japan earthquake

In 2011 Japan suffered its most powerful earthquake in a thousand years. The earthquake unleashed a tsunami up to 30 metres in height and resulted in the worst nuclear disaster in 25 years. This chapter explains why.

Japan is a small country in the Pacific Ocean in East Asia. It is made up of several islands, with Honshu, Hokkaido, Kyushu and Shikoku being the four largest (see Figure 14.1).

Despite Japan's small land area, it has the tenth largest population in the world with over 127 million people. Nearly three-quarters of these people are squeezed into a narrow coastal strip because inland is very mountainous. Over 30 million people live in just one city, Tokyo, the largest urban area in the world.

# THE CAUSE OF THE JAPAN EARTHQUAKE, 2011

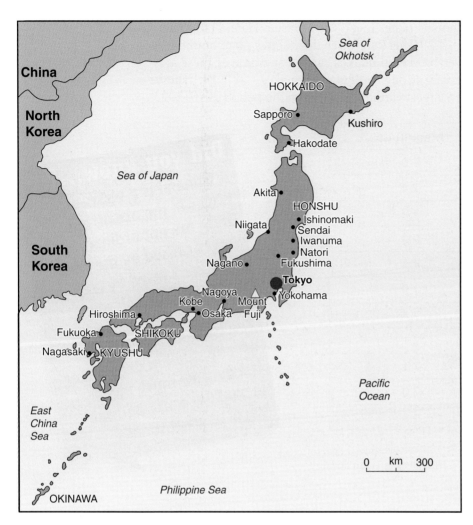

**Figure 14.1**
Japan

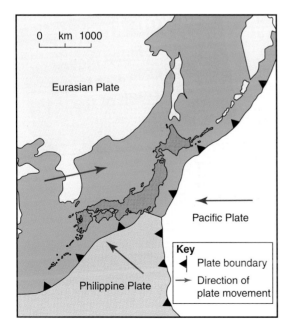

**Figure 14.2**
Crustal plates near Japan

Japan lies in an area where three crustal plates meet: the Eurasian Plate, Philippine Plate and Pacific Plate (see Figure 14.2). These are three destructive plate boundaries and, for this reason, Japan has over 1000 earthquakes each year. Some of these are only tremors; however, others are very strong, violent earthquakes. Table 14.1 shows the dates and magnitude of the ten largest earthquakes during the period 2010–2012.

| When? | Magnitude |
| --- | --- |
| 26 February 2010 | 7.0 |
| 21 December 2010 | 7.4 |
| 9 March 2011 | 7.2 |
| 11 March 2011 2:46 p.m. | 9.0 |
| 11 March 2011 3:08 p.m. | 7.4 |
| 11 March 2011 3:15 p.m. | 7.9 |
| 11 March 2011 3:25 p.m. | 7.4 |
| 7 April 2011 | 7.1 |
| 11 April 2011 | 7.1 |
| 10 July 2011 | 7.0 |
| 7 December 2012 | 7.3 |

**Table 14.1**
Dates and magnitude of the ten largest earthquakes in Japan during 2010–2012

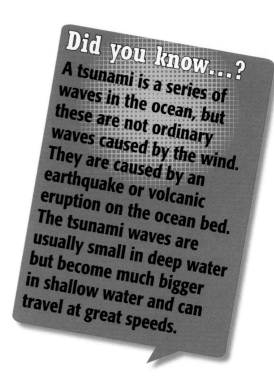

**Did you know...?** A tsunami is a series of waves in the ocean, but these are not ordinary waves caused by the wind. They are caused by an earthquake or volcanic eruption on the ocean bed. The tsunami waves are usually small in deep water but become much bigger in shallow water and can travel at great speeds.

## Cause of the 2011 earthquake

At 2:46 p.m. **on 11 March 2011, a magnitude 9.0 earthquake hit the northeast coast of Japan**; its epicentre was 70 km offshore and the focus was 6 km below the ocean bed. **The earthquake was caused by the Pacific Plate being pushed under the Eurasian Plate.** It moved 20–40 metres. The sudden jolting of the plates created the initial earthquake and was followed by more than 50 aftershocks greater than magnitude 6.0. The earthquake itself lasted for over five minutes and **the sudden uplift of the sea floor caused tsunami waves to spread out across the ocean at speeds of up to 800 km/h**. Tsunami warnings were issued as far away as Hawaii, Australia, Fiji, Mexico and Chile. The highest tsunami waves were over 30 metres high.

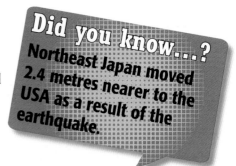

**Did you know...?** Northeast Japan moved 2.4 metres nearer to the USA as a result of the earthquake.

## Predicting the earthquake

Earthquakes are one of the most difficult natural disasters to predict but progress has been made in recent years. **Scientists had expected an earthquake of this magnitude in Japan for the following reasons:**

- Japan lies on the boundary of three crustal plates – all of which form destructive plate boundaries.

# THE CAUSE OF THE JAPAN EARTHQUAKE, 2011

- They are also 'active' plate boundaries, with over 1000 earthquakes occurring there each year.
- In particular, in the previous week there had been several substantial earthquakes, one measuring 7.2 on the Richter scale.

Although people knew that a serious earthquake would happen in the near future, **experts could not predict exactly where or exactly when**. Scientists do not yet know when rocks that are under pressure at a plate boundary will suddenly move. For this earthquake, **seismometers detected strong movements in the crust one minute before the earthquake struck** which allowed warnings to be sent all over Japan immediately.

Although a big earthquake releases pressure at one part of the plate boundary, this pressure is then transferred to another part nearby, which increases the chances of another earthquake there in the near future.

## National 4

1. Why does Japan experience so many earthquakes?
2. Look at Table 14.1 and rank the earthquakes in order of strength.
3. According to Table 14.1, which year had the most earthquakes? How many occurred?
4. Describe the reasons for the Japanese earthquake on 11 March 2011.
5. What is a tsunami?
6. What caused the tsunami on 11 March 2011?
7. Was it possible to predict the earthquake on 11 March 2011? Explain your answer.

## National 5

1. Explain why Japan experiences so many severe earthquakes.
2. Using Table 14.1, rank the earthquakes in order of strength.
3. (a) Explain in detail why the Japanese earthquake on 11 March 2011 occurred.
   (b) On that day there were three more major earthquakes in that area. Why do you think that was?
4. Explain why a tsunami followed the earthquake.
5. Explain fully why it is so difficult to predict earthquakes.

## Activities

### Activity A

Carefully read the story below. You will notice that the story ends suddenly. You have to think of an appropriate ending for the story.

At 2:40p.m. on 11 March I was at school, sitting at my desk with all my classmates. We were in Geography learning about population distribution. I was listening carefully to my teacher as I really enjoy Geography so I always pay attention. A few minutes later, out of nowhere, the ground started to shake, I could feel and hear my desk rattling and the windows in the classroom started to rattle too. Some of my classmates started to scream and we soon realised what was happening. My teacher tried to keep calm but I could tell by the expression on her face that she was panicking too. She shouted at us all to get under our desks. The walls were shaking and creaking and I could hear bits of the ceiling fall off and hit the floor and desks. The lights started to flicker and suddenly everything went quite dark …

### Activity B

In each question below, there are two sentences describing two events connected with the earthquake. Copy the sentences carefully into your notebook and decide whether or not there is a connection between them. Next to each question write one of three letters:

- **M** if there *must* be a connection between the two.
- **C** if there *could* be a connection.
- **N** if there *cannot* be a connection between the two events and they just happened to occur at the same time.

1. Japan is made up of several small islands. Japan has a population of 127 million.
2. Japan lies near three destructive plate boundaries. Japan has many earthquakes and volcanoes.
3. Japan has many mountains and volcanoes. Japan has a very high population density.
4. Japan has many earthquakes. Japan has many tsunamis.
5. Japan experienced a very strong magnitude 9.0 earthquake. Japan lies near three destructive plate boundaries.
6. Japan gets over 1000 earthquakes each year. Predicting exactly when earthquakes will happen is impossible.

**Now complete the 'I can do' boxes for this chapter.**

# Chapter 15

*This chapter looks at the impact of the 2011 Japan earthquake.*

## The effects and management of the Japan earthquake, 2011

**By the end of this chapter, you should be able to:**

- ✓ describe the impact of the earthquake on the landscape
- ✓ give examples of how the earthquake affected the people of Japan
- ✓ describe how Japan coped with this earthquake.

## Impact on the landscape

The earthquake that hit Japan on 11 March 2011 released the same amount of energy as two million atomic bombs. The P waves travelled from the epicentre at 6 km/sec and the more destructive S waves followed at 3 km/sec. Within one minute of the earthquake, northeast Japan was shaking violently. **Buildings and bridges collapsed.** The shaking ground caused **landslides which buried cars, destroyed homes and blocked roads**. Electricity cables and gas pipes were badly damaged causing **thousands of homes and buildings to catch fire** and several oil refineries went up in flames. The earthquake was so violent that much of **the east coast of Japan dropped by one metre**.

The impact of the earthquake was very serious but worse was to come. The first tsunami waves hit the northeast coast 20 minutes after the earthquake. **The tsunami had waves of up to ten metres** crashing against the coastal towns. Tsunamis have devastating power because they are waves of debris not just water. Within the waves are cars, boats, whole houses and buildings and all their contents. **As the tsunami swept 10 km inland it demolished everything in its path – houses, farms and factories.**

In total, **2000 km of coastline was devastated** by the earthquake and the tsunami. In cities such as Iwanuma and Natori, homes, businesses and communications were completely obliterated by the waves (see Figures 15.1 and 15.2). **Several towns and villages were wiped off the map.** Altogether, at least **1.2 million buildings were damaged, destroyed, washed away or burnt down.**

**Figure 15.1**
Tsunami waves hitting Iwanuma, Japan

**Figure 15.2**
Tsunami waves hitting Natori, Japan

Meanwhile, the most serious effect of the earthquake was only just beginning. When the earthquake struck, the six reactors which make up Fukushima nuclear power plant automatically shut down but the cores of the reactors were still extremely hot and had to be cooled.

**Did you know...?** The tsunami left 29 million cubic metres of waste – enough to fill Wembley Stadium 25 times.

The earthquake caused power cuts which stopped the cooling systems from working. Emergency generators cooled them but these were then flooded by the tsunami. The reactors continued to heat up and Reactors 1, 2 and 3 experienced full meltdown, which led to a number of chemical explosions. For many days it was feared that there would be a complete meltdown with large amounts of radioactive material being released into the atmosphere. However, sea water was finally used to cool down the reactors and save the plant, although it will never be used again.

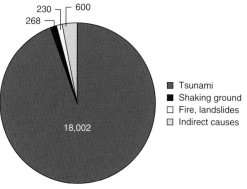

**Figure 15.3**
Numbers killed in the 2011 Japan earthquake

## Impact on people

A few of the nuclear power plant's workers were severely injured or killed by the disaster. Some were exposed to high levels of radiation and **it has increased the risk of cancer in local residents by about 1%**. Infants who were exposed to the radiation have the highest risk.

Altogether, **the earthquake and tsunami killed approximately 19,000 people** (see Figure 15.3). Ishinomaki city had the most recorded fatalities with 3735. People over age 65 made up 56% of the deaths.

The government closed down eleven other nuclear reactors for safety reasons, causing **power shortages** for several days. These affected the whole economy. In total **500,000 people were made homeless** by the earthquake and were in serious danger of hypothermia as temperatures plunged to –5 °C, with freezing winds, hail storms and thick snow affecting the country. People were forced to scavenge for food in the debris, as there were **major food shortages** across the affected areas. Four million people were without water and six million without electricity.

In total, the cost of the earthquake was as much as $300 billion, which makes it the costliest natural disaster ever and brought an economic crisis to Japan.

## Planning for the earthquake

Japan is better prepared than any other country for earthquakes.

- Japan has an earthquake detection system and this picked up the earthquake as it happened. Automatic warnings were sent across the country within 31 seconds; one even interrupted Parliament.
- People in Japan know what to do in the event of an earthquake. Children have regular earthquake drills in school; offices, houses and schools all have emergency earthquake kits.
- There is a tsunami warning centre in Hawaii and a tsunami warning was sent to Japan 10 seconds after the earthquake. People were told to head for higher ground but in many areas this was too far away.
- There are tsunami drills so people know what to do.
- Coastal towns had already built sea walls up to 10 m high to keep out tsunami waves but, in some cases, these were not high enough because the earthquake had caused the land and the defence walls to sink.

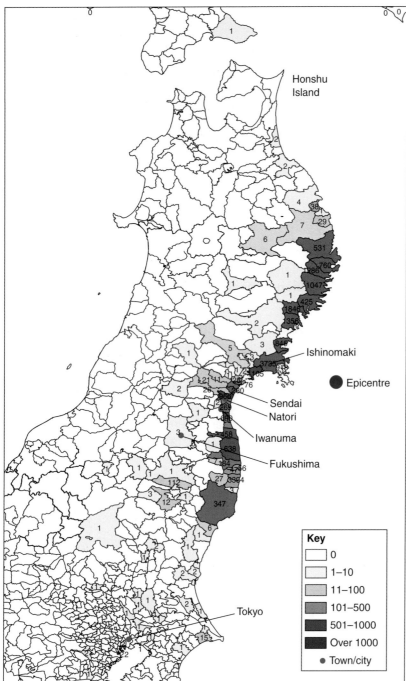

**Figure 15.4**
Numbers killed in the 2011 Japan earthquake

- Japan's planning measures saved many lives. Buildings are regularly checked to make sure they sway rather than shake during an earthquake and the majority of the buildings survived this earthquake because they were 'earthquake-proof'. The early warning earthquake and tsunami system worked. People knew of the earthquake immediately and they knew what to do when an earthquake occurs. Despite all these measures, 19,000 people died.

THE EFFECTS AND MANAGEMENT OF THE JAPAN EARTHQUAKE, 2011    83

# The relief effort

Immediately after the earthquake and tsunami struck, local people and charities started to provide assistance. The local authorities and the national government then stepped in and were quickly followed by promises of aid from other countries. Figure 15.5 shows examples of some of the different types of aid given by selected countries.

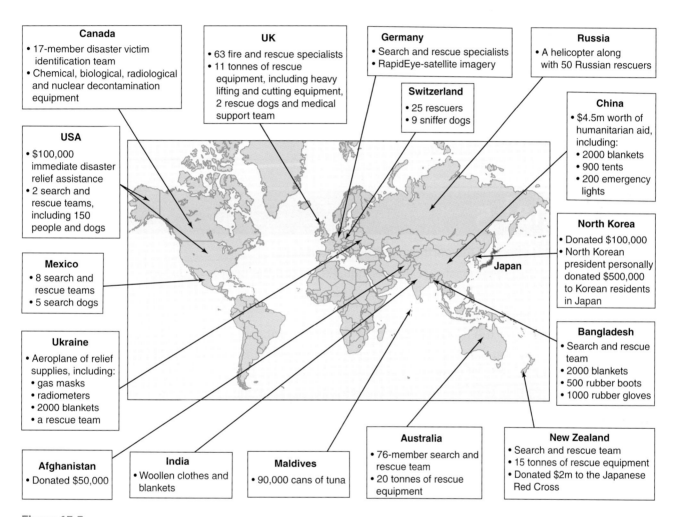

**Figure 15.5**
Examples of international aid given to Japan

# National 4

| Impact on the landscape | Impact on people |
|---|---|
|  |  |

1. Make a table in your notebook like the one above. Use the information in this chapter to complete your table.
2. Look at Figure 15.3. Which was more deadly – the earthquake or the tsunami? Explain why.
3. Why do you think so many older people died in the earthquake?
4. Japan was prepared for a severe earthquake. Describe two of the planning measures and explain how they saved lives.
5. Although Japan was well-prepared, many lives were lost. Explain why.
6. Using Figure 15.5, in your opinion what were the two most important types of aid that were given? Why?

# National 5

1. Summarise the effect of the earthquake on the landscape.
2. Describe, in detail, the impact that the earthquake had on the people of Japan.
3. Look at Figure 15.3.
   (a) Describe what is shown.
   (b) Give reasons for the different causes of deaths shown.
4. Some people say that Japan's planning for this earthquake was very effective; others say it was not. Explain both points of view.
5. Using Figure 15.5, in your opinion what were the three most important types of aid that were given? Why?

# Activity

Design and write a newspaper front page for the day after the Japan earthquake. You should:
- give your newspaper a title at the top and a date
- have an eye-catching headline underneath.

Now divide the rest of your page into two or three columns and have between two and four sections in each column. In each section write about one aspect of the earthquake, such as:

- where and when it happened
- its cause
- its strength
- the effects of the earthquake
- the tsunami
- the relief effort.

If possible, include a picture and a map.

**Now complete the 'I can do' boxes for this chapter.**

# Chapter 16

## Tropical storms

*This chapter looks at the location and features of tropical storms.*

**By the end of this chapter, you should be able to:**

- ✓ describe what a tropical storm is
- ✓ describe the main features and locations of tropical storms
- ✓ describe the conditions needed to create a tropical storm.

## Tropical storms

**Tropical storms are severe depressions** in which **wind speeds reach over 60 km/h** but can often reach over 200 km/h. As Figure 16.1 shows, tropical storms are **found over oceans within 30 degrees of the equator**. They start on the eastern side of oceans and move westwards, before dying out over land. **When tropical storms reach 120 km/h, they are called hurricanes.** There are local names for hurricanes in different parts of the world, as shown in Figure 16.3 on page 88.

## Main features of tropical storms

About 500 million people in 50 countries live in fear of tropical storms. They kill more people each year than earthquakes or volcanoes, yet some parts of a tropical storm are much more deadly than others. The main features of a tropical storm are described below and shown in Figure 16.1.

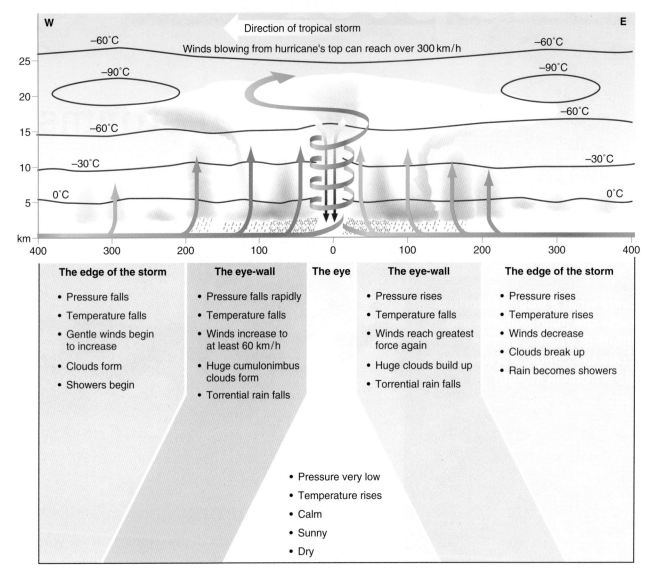

**Figure 16.1**
Main features of a tropical storm

1. As the storm approaches, the air pressure and temperature drop, while cloud cover and rainfall increase.
2. **Near the centre, at the eye-wall**, huge cumulonimbus clouds rise up, **torrential rain falls and wind speeds reach their maximum**.
3. **At the centre, the eye is calm, clear, warm and dry**.
4. After the centre is the other eye-wall and the same weather as in point 2 is experienced again, with towering clouds, very heavy rain and very strong winds.
5. At the edge of the storm, the air pressure and temperature rise, while cloud cover and rainfall decrease.

**A tropical storm travels at about 10 km/h**, but it can speed up or slow down quickly. The route a tropical storm takes is called a 'track' and **it can change direction suddenly**. On reaching coastal areas, **it can raise the level of the surface water** by up to ten metres. At high tides **this produces a storm surge**, which leads to severe flooding. Once a tropical storm reaches land it slows down, changes direction and quickly dies out. An average tropical storm lasts for one to two weeks.

# TROPICAL STORMS

## Conditions needed for a tropical storm

Tropical storms are only found in certain areas of the world, as shown in Figure 16.3. These are the areas that have the necessary conditions for them to form. A tropical storm needs the following conditions to form:

1. **Warm seas, which have a surface temperature of 27 °C or more**, and warm water to a depth of at least 60 metres.
2. **Low air pressure**, with the air beginning to rise.
3. **Damp moist air** with a relative humidity of 60% or more.

Where these conditions are found, there are five stages in the formation of a tropical storm. These are shown in Figure 16.2.

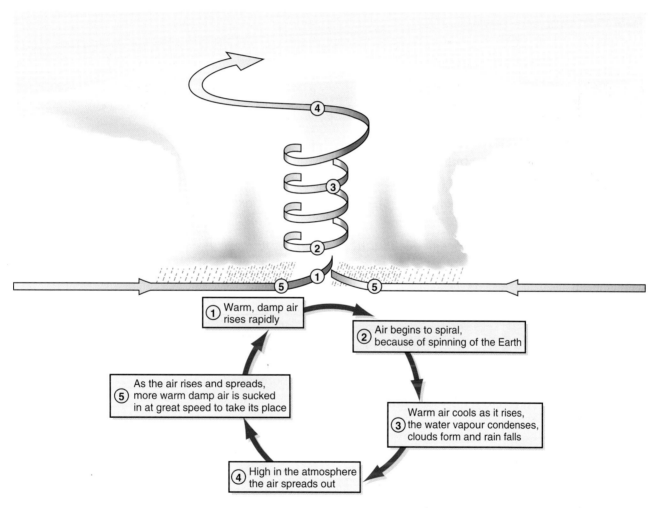

**Figure 16.2**
Stages in the formation of a tropical storm

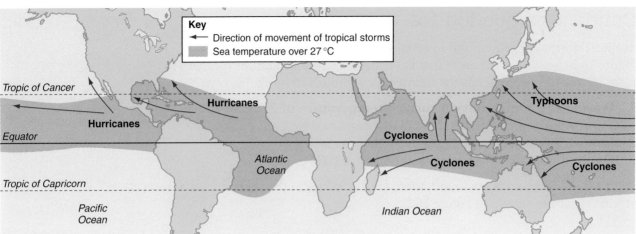

**Figure 16.3**
Distribution of tropical storms

## National 4

1. What are tropical storms?
2. Describe the distribution of tropical storms in the world.
3. Describe the weather conditions at the eye-wall.
4. Where is the calmest weather found? Describe the weather experienced there.
5. What is a storm surge?
6. What are the three conditions needed for tropical storms to form?
7. What are the five stages in the formation of a tropical storm?

## National 5

1. What is the difference between a tropical storm and a hurricane?
2. What name is given to a tropical storm in the
    (a) Atlantic Ocean
    (b) Indian Ocean
    (c) Pacific Ocean?
3. Which part of a tropical storm brings the worst weather? Describe, in detail, the weather it brings.
4. Where is the calmest weather found? Describe the weather experienced there.
5. Describe the movement of tropical storms.
6. Describe the conditions needed for tropical storms to form and explain why these conditions are needed.
7. Why does rapidly rising air lead to
    (a) heavy rain
    (b) very strong winds?

# TROPICAL STORMS

## Activities

### Activity A

Using an atlas, name each of the countries below. Then, using Figure 16.3 to help you, decide whether or not these countries experience tropical storms.

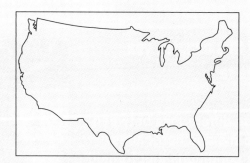

### Activity B

Carefully read the statements below. You need to decide whether the statements are true or false. If the statements are false, you must correct the information to make them true.

1. Tropical storms are found within 40 degrees of the equator.
2. When tropical storms reach 120 km/h they are called hurricanes.
3. Wind speeds become stronger towards the centre of a storm.
4. Tropical storms travel at approximately 60 km/h.
5. Once tropical storms reach land they speed up.
6. Tropical storms need seas over 27 °C, warm water to a depth of 600 metres, high pressure and damp moist air.

**Now complete the 'I can do' boxes for this chapter.**

# Chapter 17

## The cause of Hurricane Irma, 2017

*This chapter looks at the causes of Hurricane Irma.*

**By the end of this chapter, you should be able to:**

- ✓ describe the differences between tropical storms and hurricanes
- ✓ explain the conditions that formed Hurricane Irma
- ✓ describe the path of Hurricane Irma.

## Hurricane Irma

Hurricane Irma made landfall as a Category 5 hurricane on the Caribbean island of Barbuda on 4 September 2017, before laying waste other Caribbean islands and heading for the USA. It was one of the most intense hurricanes to have affected the USA since Hurricane Katrina in 2005. It caused catastrophic damage to both the USA and the Caribbean and is recognised as one of the worst natural disasters in history.

Hurricane Irma was the second major hurricane recorded in the Atlantic in 2017, following Hurricane Harvey in August, but it was the first Category 5 hurricane of that year. The cause of Hurricane Irma is shown in Figure 17.1 and its timescale is described below. Figure 17.2 shows the path and strength of the hurricane.

# THE CAUSE OF HURRICANE IRMA, 2017

## 26 August 2017

Irma began as a tropical wave over western Africa on 26 August.

After a hot summer, the sea in this part of the Atlantic Ocean reached 27°C.

Water vapour evaporated from the sea, making the air above humid and damp.

## 28 August 2017

The air in contact with the sea became very hot and began to rise, creating low pressure.

Air was then sucked in over the sea to replace the rising air. A tropical depression had now formed. This was the birth of Hurricane Irma.

As the humid air rose and cooled, the water vapour condensed, huge cumulonimbus clouds built up, thunderstorms developed and torrential rain began to fall.

## 30 August 2017

By now, the hot air was rising even more rapidly, so air rushed in even faster, making stronger and stronger winds.

The wind speeds reached 39mph and Irma became a tropical storm.

Irma started to move west towards the Caribbean.

## 31 August 2017

Wind speeds were increasing rapidly and Irma became a Category 1 hurricane when wind speeds reached 74mph.

Wind speeds continued to increase rapidly; within 12 hours, Irma had grown into a Category 3 hurricane.

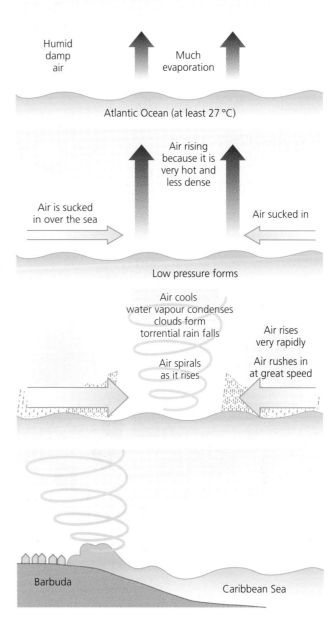

**Figure 17.1**
Formation of Hurricane Irma

## 4 September 2017

As Irma rapidly moved across the Atlantic Ocean, it intensified into a Category 4 hurricane, with wind speeds recorded as 152mph.

## 5 September 2017

By 11:45 a.m. Irma had intensified to a Category 5 hurricane, with wind speeds of 175mph.

### 6 September 2017

Hurricane Irma eventually made landfall on the night of 6 September, when it hit the island of Barbuda.

Hurricane Irma was at its strongest here, with wind speeds recorded as 185mph when it hit land.

Irma continued as a Category 5 hurricane and went on to destroy St Martin, St Barts, Anguilla, the British Virgin Islands, the Dominican Republic and Haiti.

### 8 September 2017

Irma made landfall as a Category 5 hurricane in Cuba. This was Cuba's first Category 5 hurricane in decades.

### 10 September 2017

Hurricane Irma reached the Florida Keys as a Category 3 hurricane. Once over land, without any moist air, Irma lost energy and began to slow down.

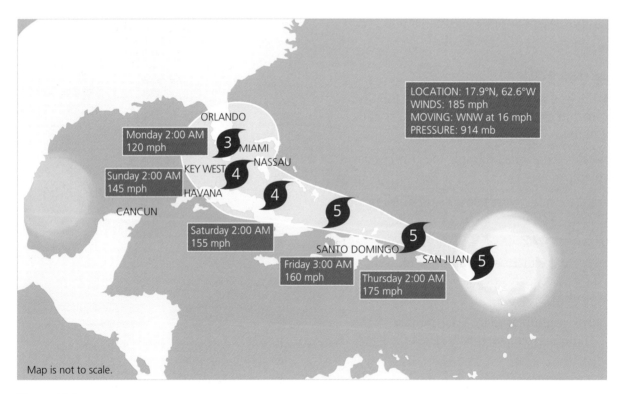

**Figure 17.2**
The path and strength of Hurricane Irma

# THE CAUSE OF HURRICANE IRMA, 2017

## National 4

1. Which countries were affected by Hurricane Irma?
2. Explain why air began to rise over the Caribbean at the end of August.
3. Explain how rising air led to the very strong winds of Hurricane Irma.
4. At what wind speed did Hurricane Irma become:
   (a) a tropical storm
   (b) a hurricane?
5. Using Figure 17.2, describe the path of Hurricane Irma.
6. Where did Hurricane Irma eventually lose energy on 10 September, and why?
7. Copy the table below and complete it to show the movement of Hurricane Irma between 31 August and 10 September.

| Date | Location of Irma | Wind speed |
|---|---|---|
| 31 August | | |
| 4 September | | |
| 5 September | | |
| 6 September | | |
| 8 September | | |
| 10 September | | |

Table 17.1

## National 5

1. Describe the region affected by Hurricane Irma, including the names of states, islands and seas.
2. Explain why Hurricane Irma:
   (a) had such strong winds
   (b) brought heavy rain.
3. Using Figure 17.2, describe the path of Hurricane Irma.
4. Copy Table 17.1 and complete it to show the movement of Hurricane Irma between 31 August and 10 September 2017.
5. Explain the changing wind speed of Hurricane Irma between 31 August and 10 September.

# Activities

## Activity A

The statements below explain the formation of a hurricane. Rearrange the statements in the correct order.

- When wind speeds reach 60 km/h a tropical storm is formed.
- The air becomes very hot and starts to rise, creating low pressure.
- Evaporation occurs in seas where the temperature is over 27 °C.
- More and more air rushes in, which makes the winds become stronger and stronger.
- When wind speeds reach 120 km/h a hurricane forms.
- Air is sucked in under the rising air to create a tropical depression.

## Activity B

Table 17.2 below shows the Saffir Simpson scale, which is used to measure the strength of a hurricane to produce a category numbered from 1–5. Using this, decide what category of hurricane produced the damage in pictures A, B and C. Explain your decision.

| Category | Wind speeds (km/h) | Effects |
| --- | --- | --- |
| 1 | 120–155 | Trees and power lines can be brought down. Damage to buildings is minimal. Flooding might occur. |
| 2 | 155–180 | Some structural damage to buildings. Widespread damage to trees and farmland. Power lines will be downed. Widespread flooding. |
| 3 | 180–210 | Significant damage to buildings. Mobile homes will be completely destroyed. Storm surges may cause extreme flooding. |
| 4 | 210–250 | Extensive damage to buildings. Major damage to area. Storm surges may cause extreme flooding. |
| 5 | More than 250 | Complete devastation of buildings. Major roads destroyed or cut off. Vegetation is completely destroyed. |

Table 17.2

## Activities continued...

A

B

C

Now complete the 'I can do' boxes for this chapter.

# Chapter 18

## The impact and management of Hurricane Irma, 2017

*This chapter looks at the damage caused by Hurricane Irma.*

**By the end of this chapter, you should be able to:**

- ✓ explain the impact of Hurricane Irma on the landscape
- ✓ give examples of the impact of Hurricane Irma on the people
- ✓ describe the relief effort after Hurricane Irma.

## Impact on the landscape

Hurricane Irma devastated many countries between 31 August and 10 September 2017. In total 1.8 million square kilometres were affected. The worst affected countries were those in the Caribbean where much of the damage was caused by the Category 5 wind speeds, especially Barbuda, St Martin and Cuba.

### Barbuda

When Hurricane Irma made landfall on the small Caribbean island of Barbuda, it was a Category 5 hurricane. In a short time, 95% of all the island's buildings were damaged by its fierce winds and much of the land was flooded by the torrential rain and lay under water for weeks. All 1800 residents were evacuated immediately after Irma to the island of Antigua for fear that they would be hit again by the next hurricane, Hurricane Jose. The island was deserted.

## St Martin

Irma was still a Category 5 hurricane when she hit the small Caribbean island of St Martin. The winds, still exceeding 150mph, took out approximately two-thirds of all the buildings on the island. Roads were submerged under water and floods carried away cars, boats and debris. The island's people were left with no electricity, gas or safe drinking water. Trees were ripped out of the ground by the devastating Category 5 wind speeds and the total damage was in excess of £1 billion.

## Cuba

Storm surges and 16-metre high waves caused by the excessive wind speeds left much of northern Cuba completely submerged under floodwater. Villages and communities were completely flooded and the people left without shelter. It is thought that over 158,000 homes were damaged by the hurricane, with approximately 15,000 homes completely destroyed. The popular tourist resorts of Cayo Coco and Cayo Santa Maria were flooded. Trees and telegraph poles were damaged and in Cayo Coco several hundred flamingos were killed. Extensive flooding was reported all over the island and the total cost of damage to Cuba was £370 million.

**Figure 18.1**
Large sections of the Caribbean were damaged in Hurricane Irma

## Florida

By the time Hurricane Irma made landfall in Florida on 10 September it had been downgraded to a Category 3 hurricane. Florida was one of the worst affected areas nonetheless. The storm surges that affected areas such as Naples and Fort Myers were as high as 4 metres, levees burst and 32 rivers overflowed, resulting in widespread flooding. Approximately 65,000 buildings were destroyed during the hurricane. It is estimated that one out of every four homes was destroyed in parts of Florida following Hurricane Irma. 12 million people were without electricity and $2.5 billion worth of damage was done to the agricultural industry.

# Impact on people

It is thought that Hurricane Irma killed 134 people in total, with the most deaths occurring in Florida, where approximately 72 people were killed. Most people drowned in the floodwater but injuries from falling buildings and carbon monoxide poisoning were also to blame.

Millions of people were told to evacuate their homes and islands, including 6.4 million Floridians, 5000 people from the Bahamas and 1700 from St Martin. 16,000 flights were cancelled to and from the Caribbean and the USA and 51 cruises had to be cancelled due to the storm.

10,000 people in Florida were made homeless following Irma, while many Caribbean nations were completely desolate following the storm. All 1800 residents of Barbuda were evacuated to Antigua after the storm and months later very few had returned as the country had been almost completely destroyed.

In total, it is thought that £38 billion worth of damage was caused by the hurricane.

| | |
|---|---|
| **Hundreds of weather stations** on land and at sea record the weather as the hurricane approaches and passes over, giving information on its windspeed, wind direction, temperature and pressure | |
| **Radiosonde balloons** are sent into the hurricane carrying weather instruments and they send back information on temperature, pressure and humidity | |
| **Radar** is used to find out where the rain is falling and its intensity | |
| **Satellites** take photographs of the hurricane so that its speed and direction can be tracked | |
| **Specially designed aircraft** fly into hurricanes and record windspeed, wind direction and temperature | |
| **Computers** in the National Hurricane Center in Miami, USA process all these data and, based on how previous hurricanes have behaved (stored in their memory), they predict the hurricane's speed, strength and direction over the next few days | |

**Figure 18.2**
Methods of forecasting hurricanes

# Planning for the hurricane

We all know how difficult it is for weather experts to forecast our weather. Hurricanes are even more difficult to predict than the weather systems which cross the British Isles and a lot of sophisticated equipment is used to predict them, especially in the USA (see Figure 18.2).

Everyone in the southern USA and the Caribbean knows the likely time of year that hurricanes strike – commonly referred to as 'hurricane season' – because they understand what causes them. The government regularly informs people of how they should prepare for hurricanes; they even have a National Hurricane Preparedness Week before the hurricane season starts. But they know that hurricanes are difficult to predict and plan for because they often change direction and speed.

The USA has a National Hurricane Center in Florida, which tracks all tropical storms. Special aircraft can fly into hurricanes to obtain weather data. The National Hurricane Center accurately predicted the path and speed of Hurricane Irma. This allowed people in all the countries in Irma's path to make preparations. Many people chose to evacuate the area before the hurricane hit, while others elected to stay and sought safety in hurricane shelters. People were told to pack bags with essential items such as water, food and first aid kits.

# Aid given following the hurricane

Hurricane Irma was one of, if not the, most powerful storm to hit the Atlantic Ocean. People affected saw their homes and livelihoods completely destroyed and urgently needed help following the storm.

Emergency aid was needed immediately after Hurricane Irma hit. Many countries all over the world donated emergency aid to help in the relief effort. The UK gave £32 million in emergency aid, providing Marines, Royal Engineers and RAF helicopters to help rescue people from floodwaters and trapped buildings. The Netherlands also sent planes with military support. France sent 140 tonnes of electrical equipment including generators and pumps.

In order to help the survivors, the Netherlands also sent food and water to the affected areas. Thousands of people were made homeless by Irma, therefore beds and blankets were urgently needed. The Red Cross distributed blankets, tarpaulins and other emergency supplies to the affected countries. More than 750 health workers were sent to the Caribbean islands, many from Cuba – itself an affected country.

In order to allow the countries affected to start rebuilding, long-term aid was needed. Because so much was destroyed by the hurricane, everything needed to be rebuilt, particularly in the Caribbean countries of Barbuda and St Martin that were almost completely laid to waste. Roads, schools, houses and hospitals all needed to be rebuilt. The UK, France and Canada all pledged to help the Caribbean countries get back on track.

## National 4

1. Describe the damage to the landscape in Barbuda, St Martin, Cuba and Florida caused by Hurricane Irma.
2. Which country do you think was most affected?
3. How many people died as a result of Hurricane Irma and what caused the most deaths?
4. Choose two methods of forecasting hurricanes and describe them.
5. How did people plan for Irma?
6. Hurricane Irma was well-predicted. Describe what might have happened if the region had been completely unprepared for it.
7. Explain why short-term and long-term aid was needed after the hurricane.
8. Do you think the aid effort was successful? Give reasons for your answer.

## National 5

1. Compare the damage to the landscape in Barbuda, St Martin, Cuba and Florida.
2. Describe the effects of Hurricane Irma on the people of the Caribbean and USA.
3. Look at Figure 18.2. Which three methods of forecasting Hurricane Irma do you think were the most useful? Explain your decision in detail.
4. The American and Caribbean people knew a severe hurricane was about to hit. Explain fully how this helped to reduce deaths and damage.

## National 5 continued...

5. Describe the types of short-term aid given and explain in detail the reasons why they were needed.
6. What is long-term aid used for and why is this important?
7. How effective was the short-term and long-term aid effort?

## Activities

### Activity A

Use a map of the south of the USA and the Caribbean, similar to the one here. Annotate the map to show the damage to each of Barbuda, St Martin, Cuba and Florida caused by Hurricane Irma.

**Figure 18.3**
Map of the south of the USA and the Caribbean

Annotate your map with the following information:

- The locations of Barbuda, St Martin, Cuba, Florida
- The path of Hurricane Irma (using Figure 17.2)
- Draw four boxes – one each for Barbuda, St Martin, Cuba and Florida – and inside each create a bulleted list of the damage caused by Irma.

### Activity B

Design a poster for a charity to persuade people to help those affected by Hurricane Irma.

Now complete the 'I can do' boxes for this chapter.

# N5 Examination questions

Read the information and the two SQA-style questions below, and answer Tasks A–H.

## Question 1

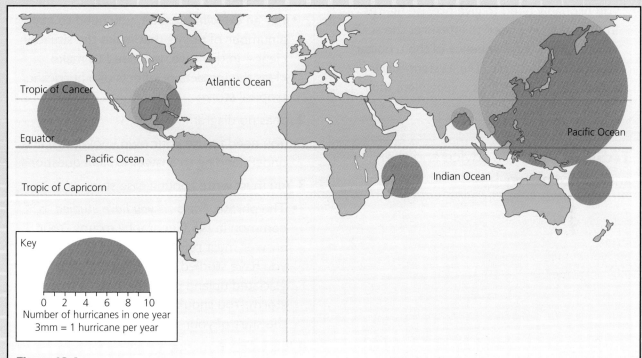

**Figure 18.4**
Diagram Q1 – Distribution and frequency of hurricanes

(a) Study Diagram Q1. Describe the distribution and frequency of hurricanes. (4)

(b) For a hurricane or tropical storm you have studied, explain its impacts on the people and landscape. (6)

### Advice for Question 1(a)

You should have spotted four important points.

1 It is a *describe* question.
2 It has a diagram. For the topics in the Global Issues section of the examination, the first part of the question usually asks you to describe a graph or map.
3 There are two parts to the question. You need to describe both the distribution (where the storms are found, e.g. they occur over oceans) and their frequency (how often they occur, e.g. most occur in the Pacific Ocean) in order to earn full marks.

### Advice for Question 1(b)

You should have spotted four important points.

1 It is an *explain* question.
2 It has no diagram.
3 You must write about a case study.
4 It is worth **6** marks.

### Advice for Question 1(a)

**4** It is worth **4** marks, so you need to make 4 factual points or 2 developed points. As has already been mentioned, to earn 2 marks when describing a graph or map, you probably need to give precise figures. In this case, you need to interpret the scale which shows the frequency of the tropical storms. For example, *The western Pacific Ocean has 17 hurricanes a year, more than anywhere else.* (2)

**TASK A**: Read the advice and then answer Question 1(a).

### Advice for Question 1(b)

Let's take these points separately.

**1** It is an *explain* question.
- This is one of six types of question you can be asked.
- For an *explain* question, you need to make a number of points that make the situation clear – in this case, you need to make clear the impacts or effects of a hurricane/tropical storm.

**2** It has no diagram.
- You need to use your own knowledge and understanding to answer *explain* questions.

**3** You must write about a case study.
- The phrase, *For a … you have studied,* is common in N5 Geography exams. Your answer must refer to something specific you have studied – a city, a rural area, a glaciated upland or, in this case, a tropical storm. You should name your case study at the start of your answer.

**4** It is worth **6** marks.
- There is 1 mark for each valid reason you give, but you can earn 2 marks by giving a developed reason. For example, *Typhoon Haiyan killed 6000 people in the Philippines in 2013;* (1) *some were killed by falling buildings while some died because the blocked roads meant they could not reach hospital in time.* (1)

**TASK B**: Read the advice and then answer Question 1(b).

# N5 EXAMINATION QUESTIONS

## Question 2

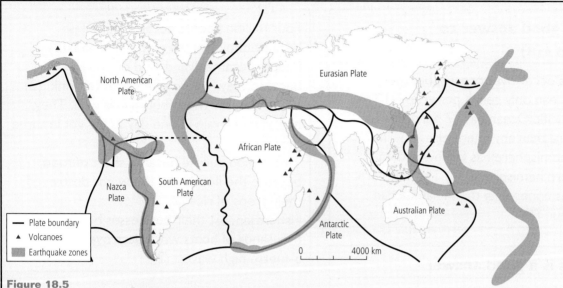

Figure 18.5
Diagram Q2 – The world's earthquake zones

(a) Look at Diagram Q2. Describe, in detail, the distribution of the world's earthquake zones. (4)

(b) Explain, in detail, the strategies used to reduce the effects of an earthquake. You must refer to named examples. (6)

### How good is this answer to Question 2(a)?

All the earthquake zones are found near crustal plate margins. This is because the plates are sliding past or under one another but become stuck. They are trying to move and so suddenly they jolt forward, causing an earthquake. Or, the plates are moving apart and the rock suddenly cracks and moves.

**TASK C:** Read the answer above and give it a mark. Explain the number of marks you have given.

### How good is this answer to Question 2(b)?

New Zealand has lots of earthquakes but they are well-prepared for them.

The people:
- are given earthquake drills and told to 'drop, cover and hold', so they are not killed by falling buildings
- are given instructions on what to do by the government, for example storing water and food – just in case they need it after the earthquake.

The government:
- has devices in the ocean giving warnings of tsunamis, so people can move to higher ground in time
- makes sure all new buildings are earthquake-proof, with deep foundations, so they only sway – they don't collapse
- is making old brick buildings stronger
- is making water and sewage pipes that bend rather than break, so people can still have clean water after an earthquake.

**TASK D:** Read the answer above and give it a mark. Explain the number of marks you have given.

## Question 1 on page 101

### This is a good answer to Question 1(a)

Hurricanes occur over oceans, but the Atlantic Ocean only gets 5 per year. (1) The western Pacific Ocean has 17 hurricanes a year, more than anywhere else. (2) The southern hemisphere has far fewer than the northern hemisphere. (1) The southern Atlantic has none while the Indian Ocean only has four. (1)

### Why this is a good answer

The answer has facts about the distribution of hurricanes **and** their frequency. For distribution, the answer refers to the overall pattern (oceans, northern hemisphere) and particular locations (Europe, Pacific Ocean). For frequency, the answer includes precise numbers worked out from the scale. With at least four good points, this answer scores full marks.

**TASK E**: The answer gives the number of hurricanes in three locations. Write down the number of hurricanes in the other locations.

### This is a good answer to Question 1(b)

<u>Typhoon Haiyan (2013), Philippines</u>
6000 people were killed by the very strong winds and the very heavy rainfall. (1)
The winds reached nearly 200mph which killed many people. Some were killed by falling buildings while others died because the blocked roads meant they could not reach hospital in time. (2)
Four million people became homeless because their houses were flattened. (2)
People became ill and died of diseases because they had no safe water to drink. (2)
It was not just winds but rain as well. They had as much rain in one day as we get in three months.
Two million people became short of food because the floods and the winds destroyed their crops of rice (2)
Fishermen lost their businesses because thousands of boats were destroyed by 5-metre high waves. (2)

### Why this is a good answer

The candidate has stated the hurricane they have chosen at the start of the answer. The candidate has explained the different effects, giving a reason why people were short of food or were killed. He/she has learned this case study very well and mentions at least 6 effects and most of them are developed points.

**TASK F**: The candidate has given 5 precise numbers for this case study (for example, *6000 people were killed*) but it can be difficult to remember so many facts in an exam. Imagine you cannot remember the precise numbers and rewrite the answer (for example, you will need to write: *A few thousand people were killed*). How many marks is your new answer worth?

# Question 2 on page 103

## Advice for Question 2(a)

You should have spotted two important points.

1. It is a *describe* question and it has a diagram – Diagram Q2. You are expected to interpret the map and state a few facts about where earthquake zones are found.
2. It is worth **4** marks, so you must make 4 valid points or 2 developed points.

## How good was the answer on p.103?

### (Mark: 1 out of 4)

The first sentence describes where earthquake zones are found, but the rest of the answer explains why. This is not what the question asks and so the candidate only scores 1 mark. They needed to describe the distribution more thoroughly, for example are all plate boundaries earthquake zones, are the zones found over land more than ocean, in higher latitudes or near the equator, near volcanoes?

**TASK G**: Read the comments above and then write an improved answer to Question 2(a).

## Advice for Question 2(b)

You should have spotted four important points.

1. It is an *explain* question, so you must make it clear how the strategies used reduce the effects of an earthquake.
2. It has no diagram. You are expected to use your own knowledge and understanding of this topic.
3. You must write about case studies, which you must name.
4. It is worth **6** marks, so it helps to try to make detailed points worth 2 marks.

## How good was the answer on p.103?

### (Mark: 6 out of 6)

This is an excellent answer, except for one or two points. You will be penalised in the exam if your answer is a list. This answer is in bullet points and it might be treated as a list. It is difficult to understand why the candidate has done this. It would have been just as quick to write this answer without the bullet points, using full stops and in sentences. The sentences could even be written as separate paragraphs; this is how the good answer to Question 1(b) was written.

Secondly, the question tells candidates to refer to named examples (plural) and only New Zealand is named here. It would be safer to mention at least one other area. For example, *The government has devices in the ocean giving warnings of tsunamis, just like Japan.*

**TASK H**: Read the comments above and then write an improved answer to Question 2(b).

# Chapter 19

## The tundra climate

*This chapter looks at the tundra climate.*

**By the end of this chapter, you should be able to:**

- ✓ describe the tundra climate
- ✓ give examples of countries with a tundra climate
- ✓ interpret a climate graph for tundra regions
- ✓ give examples of ecosystems within tundra regions.

## Tundra climates

### Location

The tundra region is an area located near the Arctic Circle in the northern hemisphere, between 60 and 75 degrees latitude. There are many countries in the world with a tundra climate, shown in Figure 19.1. These include parts of Northern Canada, Northern Russia, Greenland and the very north of Norway. There is very little tundra in the southern hemisphere because there are few land masses at this latitude.

# THE TUNDRA CLIMATE

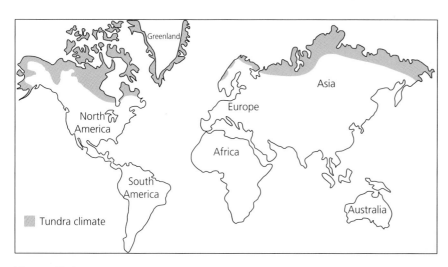

**Figure 19.1**
Location of the tundra climate

## Climate

Tundra regions are some of the coldest areas in the world. Due to their high latitudes they tend to have two seasons, winter and summer. But temperatures are low all year round, with very cold, harsh winters and cool summers. In the summer months there may be periods of continuous daylight, whereas in the winter the sun may not rise at all. Temperatures in winter range between −18°C and −50°C. In summer, temperatures remain cold and don't rise much above freezing, reaching 10–15°C. The low temperatures mean little evaporation takes place, therefore tundra regions are very dry with less than 250mm of precipitation per year. Precipitation generally falls as snow and winds can be cold and strong, making conditions very harsh.

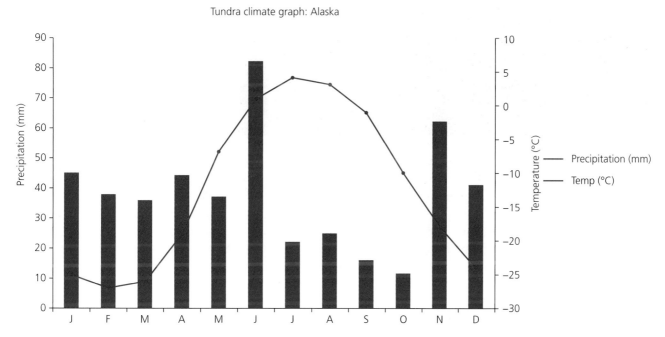

**Figure 19.2**
Climate graph of a tundra region

Figure 19.2 shows a typical climate graph for a tundra region. The graph shows both temperatures and precipitation across the year. The figures on the left-hand side of the graph show precipitation levels in millimetres and the figures on the right-hand side of the graph show temperatures in degrees Celsius. You have to be very careful when interpreting a climate graph that you are reading the information from the correct side. Precipitation is always shown as a bar graph and temperature is always shown as a line graph.

## Ecosystems

The word *tundra* means a 'treeless plain', which gives an insight into what the environment is like there. The landscape is very barren with little vegetation growing. This is ultimately due to the cold, dry climate as soil takes a very long time to form. Due to the freezing temperatures, the land is permanently frozen, known as **permafrost**. The top layers of the soil may thaw during the summer months, but the sublayers remain frozen all year round, as shown in Figure 19.3. This makes it very difficult for vegetation to grow. Plant roots can't penetrate the frozen soil therefore it is only mosses, lichens, grasses and small shrubs that can survive there. In the summer months when the permafrost melts, the water can't percolate through the impermeable permafrost below, therefore it remains on the surface creating boggy, marshy land.

**Figure 19.3**
Permafrost – sublayers of the soil remain frozen all year round

There is a very short growing season in tundra regions, between 50 and 60 days during the summer months when the sun shines 24 hours a day. Plants that grow here have to be able to withstand very difficult growing conditions. Plants must be able to tolerate freezing temperatures, they must be low-growing to survive the strong winds and must be able to survive in waterlogged conditions. They also need to have small leaves and rely on the snow to keep them insulated. Examples of vegetation that has adapted to survive in such harsh conditions include Cotton Grass, Arctic Willow and Labrador Tea.

# THE TUNDRA CLIMATE

**Figure 19.4**
Arctic Willow

Arctic Willow prefers to live in cold, dry regions. In order to survive the cold temperatures it grows long fuzzy hairs and to cope with the permafrost it has adapted to grow shallow roots and has a short growing season, all of which make it an ideal plant for the tundra.

## National 4

1. Where are tundra climates found?
2. Give three examples of countries with a tundra climate.
3. What are temperatures like in tundra regions?
4. Is there a lot of precipitation in tundra climates?
5. Why do very few plants grow in these regions?
6. What is permafrost?
7. How has the Arctic Willow adapted to the tundra climate?

## National 5

1. Describe where tundra regions are found. Use specific examples of countries in your answer.
2. Describe, in detail, the climate of the tundra.
3. Give examples of how the plants and vegetation of tundra regions are affected by the climate.

## Activities

### Activity A

| | Jan | Feb | Mar | Apr | May | Jun | Jul | Aug | Sep | Oct | Nov | Dec |
|---|---|---|---|---|---|---|---|---|---|---|---|---|
| Temperature (°C) | -22 | -20 | -20 | -16 | -10 | -2 | 4 | 8 | 0 | -9 | -14 | -18 |
| Precipitation (mm) | 5 | 4 | 4 | 5 | 9 | 10 | 25 | 35 | 25 | 20 | 10 | 5 |

**Table 19.1**
Climate in a tundra region

Using the information in Table 19.1, draw a climate graph for a tundra region.

Remember to colour your graphs appropriately.

### Activity B

Describe your climate graph in detail. You must comment on:

- Maximum temperature (figure and month)
- Minimum temperature (figure and month)
- Range of temperatures (maximum temperature minus minimum temperature)
- Wettest month (figure)
- Driest month (figure)
- Total annual rainfall.

### Activity C

Read the information below. Some of the statements are correct, while others are false. Rewrite the incorrect sentences, correcting the mistakes.

a) Tundra regions are found between 60 and 70 degrees north and south of the equator.
b) It is cold all year round in tundra regions with a very cold, harsh winter and cool summer.
c) Temperatures are above 0°C most of the year.
d) Plenty of vegetation grows in tundra regions.
e) Permanently frozen land is known as permafrost.
f) There is a very long growing season in tundra regions, between 50 and 60 days per year.

**Now complete the 'I can do' boxes for this chapter.**

# Chapter 20

*This chapter looks at the ways in which tundra environments are used.*

# How the tundra environment is used and misused

**By the end of this chapter, you should be able to:**

- ✓ give examples of ways that people use tundra landscapes to their advantage
- ✓ provide examples of ways that tundra landscapes are misused

## What is the tundra used for?

Tundra landscapes are among the harshest in the world. These unique environments create many opportunities for the people there, but can also create a number of problems.

### Hunting

Indigenous people living in the tundra use the environment in a sustainable way. This means that the practices they use do not damage the environment. Native people using these methods have lived in tundra regions for thousands of years and this is their way of life. The practices that they use include hunting caribou, seals and polar bears. They use traditional methods to fish for salmon and narwhals. Everything that is caught or killed is used; nothing is wasted.

**Figure 20.1**
Indigenous hunter

## Mining

Mining has become a profitable business in many tundra regions. Canada, Greenland and Russia all use these areas for extracting minerals such as gold, diamonds and nickel. This can be big business and creates many jobs and a lot of wealth in the region.

**Figure 20.2**
A diamond mine in a tundra region

## Oil and natural gas extraction

Alaska provides one of the best examples of oil extraction in the tundra. Oil was first discovered at Prudhoe Bay in Alaska in 1967. The oil field that was discovered was the USA's largest. The first problem the oil companies encountered was in transporting the oil. The oil needed to be transported by ship but the seas around the north coast of Alaska are frozen for most of the year. Instead they had to build a pipeline from Prudhoe Bay in Northern Alaska to the warmer waters around Valdez in the south. This Trans-Alaska Pipeline was built in 1974, with the first oil flowing through it in 1977. The 800-mile long pipeline has transported over 17 billion barrels of crude oil since its construction, earning the country billions of dollars. The construction of the pipeline created thousands of jobs and the pipeline continues to employ several hundred people in Prudhoe Bay, Anchorage, Fairbanks and Valdez.

Siberia in Russia has an abundance of natural resources. Oil and gas are extracted here, producing millions of dollars for the country. Russia has the world's largest natural gas reserves and is currently the second-largest producer of natural gas.

## New roads and settlements

The finding and extraction of natural resources from tundra climates has led to the creation of new towns to accommodate the workforce. Towns such as Anchorage and Fairbanks in Alaska have grown substantially since the discovery of oil at Prudhoe Bay. This has led to the need for all-weather roads to allow people and goods to reach them all year round.

## Tourism

Tundra landscapes offer a very unique experience for visitors. Winter snowfall and cold temperatures mean there are a number of ski resorts in tundra regions. Iceland is becoming one of the most visited countries in the world, with people visiting the tundra area to see the aurora borealis, glaciers, volcanoes and wildlife. Cruise liners are a common sight in Alaskan ports. Tourism ultimately creates many jobs through hotels, restaurants and tour operators. This feeds money into the local economy, which benefits the people living there.

## How tundra landscapes are used and misused

| Use | Positive | Negative |
| --- | --- | --- |
| Hunting | Provides a sustainable way of life for many indigenous people.<br><br>Has been practised for hundreds of years and allows people to live in an extremely harsh environment. | Tourists hunt for caribou and musk ox, not to survive but for pleasure.<br><br>This can lead to over-hunting (for example, the musk ox), which means fewer animals for the indigenous people to hunt. |
| Mining | Canada, Greenland and Russia mine gold, diamonds and nickel. This can be big business and can create many jobs in the local area and lead to many services opening up, creating a multiplier effect.<br><br>Has raised living standards.<br><br>Provides the countries with valuable exports.<br><br>Has led to more and better roads. | Pollution from mining contaminates the air, lakes and rivers.<br><br>Has led to more development, which damages the environment. |

continued

| Oil and gas extraction | In Alaska and Russia. | The building of pipelines and platforms to extract and transport oil and gas results in damage to tundra vegetation and wildlife. |
| --- | --- | --- |
| | Has created thousands of well-paid jobs, raising living standards. | Burst pipes have caused a number of oil spills over the years, resulting in catastrophic damage to the environment. |
| | Has led to more services, entertainments and roads, improving quality of life. | See case study of Alaska (Chapter 21). |
| | The companies pay taxes which provide their governments with more money to improve conditions here. | |
| | Means they do not have to import oil and gas from other countries, which is expensive. | |
| New industries, roads and settlements | The mining industry has encouraged related industries to set up, providing employment. | New towns and developments can spoil the natural landscape. |
| | New roads allow local people to get around more easily. | More roads means more vehicles, which leads to noise and air pollution. |
| | | More people means a greater threat to the very fragile vegetation. |

**Table 20.1**
Positive and negative uses of tundra landscapes

## National 4

1. Describe how indigenous people hunt in the tundra.
2. What benefits do industries such as mining bring to tundra regions?
3. Why have new roads and settlements been built in Alaska?
4. What benefits do tourists bring to the people living in the tundra?
5. What is meant by the term 'sustainable'?
6. What negative effects does new industry bring to the tundra?

## National 5

1. Describe how indigenous people hunt in the tundra.
2. Describe, in detail, the benefits that industry brings to tundra regions.
3. Give examples of the ways that tourism benefits people living in tundra regions.
4. Is hunting in the tundra truly sustainable? Give reasons for your answer.
5. In your opinion, is the tundra being misused? Give reasons for your answer.

Now complete the 'I can do' boxes for this chapter.

# Chapter 21

## Alaska: the effects of human activities

This chapter looks at the effects of human activity in Alaska.

**By the end of this chapter, you should be able to:**

✓ describe the main human activities in Alaska
✓ describe the effects that these activities are having on the people and the environment.

## Recent activity

Alaska is a state of the United States of America, separated from the rest of the country and located to the west of Canada. The majority of Alaska lies between 60°N and 70°N and, therefore, a large part of it has a tundra climate.

There has been a lot of development in Alaska since the discovery of oil in Prudhoe Bay in the 1970s. Since this discovery large-scale developments have taken place, such as the building of the Trans-Alaska Pipeline in 1976. This is an 800-mile long pipeline that transports oil from Prudhoe Bay in the north to the town of Valdez in the south.

**Figure 21.1**
Location of Alaska

The building of the pipeline was not straightforward and the engineers had to overcome a number of problems such as the large temperature ranges, the permafrost, the earthquakes and active volcanoes, as well as the mountainous terrain. All of these posed great problems to the engineers. The pipeline is not the only major recent development; there are also a number of mines in the state, extracting gold, silver and lead. These developments have undoubtedly brought many benefits to the area, but there has been a significant negative impact on the people and the environment of Alaska too.

## Impact on people

Alaska has many indigenous cultures living within the state. These indigenous people, such as the Inuits and the Yupik, have lived in this area for hundreds of years and many still retain a very traditional way of life. The building of the pipeline created many problems for these native people as much of their hunting and grazing land was lost when the pipeline was built. This meant that they had to change their practices. Many chose to abandon the traditional way of life and, instead, found employment in some of the new industries that opened up. Despite this offering the people many opportunities, it also means that their traditional practices and way of life are being lost.

**Figure 21.2**
The Trans-Alaska Pipeline

ALASKA: THE EFFECTS OF HUMAN ACTIVITIES

The creation of new industries has led to the expansion of settlements in Alaska, such as Anchorage and Fairbanks. These towns have grown in order to accommodate the growing population. Many people migrated here in order to find work in the oil and mining industries. With more people, settlements, factories and mines, there is a need for new infrastructure to be created as well. Communications have had to quickly improve in order to make these new industries and towns accessible. New and better roads have been built all over the state to allow goods and people to be transported. This inevitably leads to more vehicles, which leads to noise and air pollution. With established industries now in Alaska, other industries are being attracted to the area, such as service industries, leisure and entertainment facilities, which creates more jobs and more wealth – and more pollution.

**Figure 21.3**
Indigenous person

## Impact on the environment

Tundra landscapes are very fragile environments and even the most minor change can have serious implications for the environment. The building of the Trans-Alaska Pipeline has had a huge impact on the environment of Alaska. The 800-mile long structure has been built above and below ground. Vegetation and wildlife were damaged and destroyed in the building of the pipeline, and in areas where the pipeline is above ground it has disturbed animals' migration paths. Caribou in Alaska have had their natural migration routes disturbed and they have been forced to find new routes. This means that the native communities, such as the Inuits who hunt caribou, have also had to alter their practices.

The pipeline and oil platforms pose serious problems to the environment. Over the years there have been a number of burst pipes and oil spills that have spilt hundreds of thousands of gallons of oil on to the land and sea, devastating these fragile environments. The tankers transporting the crude oil can cause mass devastation if they suffer a spillage, like the Exxon Valdez spill in 1989. During this oil spill, 10.8 million gallons of crude oil escaped into the sea. 11,000 square miles of sea were affected, with 1300 miles of coastline coated in oil. Hundreds of thousands of seabirds, sea mammals and fish were killed in the spill.

**Figure 21.4**
Effects of oil spills

Pollution from other industries is also a huge problem in Alaska. There are a number of mines in Alaska and pollution from these is a major problem, particularly if the waste matter gets into nearby lakes and rivers.

Tundra environments do not recover quickly from any human interference. The effects of human activity are noticeable all over the country; these can be real scars on the landscape and any damage can take hundreds of years to restore.

Global warming is also affecting tundra regions at an alarming rate. It is melting the permafrost, which can lead to wide-scale flooding in some areas. Arctic permafrost contains a huge amount of carbon, approximately 1.8 trillion tonnes. When it melts, the carbon dioxide is released into the atmosphere, making global warming worse. Global warming is also causing the ice caps to melt and consequently sea levels are rising, leading to the loss of low-lying areas around the Alaskan coast.

# National 4

1. What has caused the most recent developments in Alaska?
2. What problems did engineers face when building the Trans-Alaska Pipeline?
3. How have recent developments in Alaska affected the people living there?
4. In what ways has the environment been damaged by the building of the Trans-Alaska Pipeline?
5. How is global warming affecting tundra regions?

ALASKA: THE EFFECTS OF HUMAN ACTIVITIES

# National 5

1. Give examples of recent human activity in Alaska.
2. What problems did engineers face when building the Trans-Alaska Pipeline?
3. Give examples of how recent developments in Alaska have affected the people living there.
4. Describe, in detail, the effects of the Trans-Alaska Pipeline on the environment of Alaska.
5. Explain how global warming is affecting the tundra.

## Activities

### Activity A

The worst ever oil spills are shown in the table below. Draw a bar graph to show the amount of oil that spilled during these disasters.

| Location | Year | Millions of gallons of oil spilled |
|---|---|---|
| Kuwait | 1991 | 330 |
| Gulf of Mexico | 2010 | 210 |
| Bay of Campeche, Mexico | 1979 | 140 |
| Tobago | 1979 | 88 |
| Usinsk, Russia | 1983 | 84 |
| Iran | 1983 | 80 |
| Angola | 1991 | 80 |
| Brittany, France | 1978 | 69 |
| Italy | 1991 | 45 |
| Scilly Isles, UK | 1967 | 36 |

Table 21.1
The worst oil spills around the world

### Activity B

Find the latitude of the 10 places in Activity A and work out which of these oil disasters happened in the tundra.

## Activities continued...

### Activity C

**Figure 21.5**
The Exxon Valdez oil spill in 1989

Research the Exxon Valdez oil spill that occurred in 1989. Create a report highlighting:

- When it happened
- Why it happened
- Damage that was caused to animals
- Damage that was caused to the environment
- How the area was 'cleaned up'
- Long-lasting damage.

Write your report in the style of a newspaper article, informing people of the incident.

**Now complete the 'I can do' boxes for this chapter.**

# Chapter 22

## The management of human activities in the tundra

*This chapter looks at the strategies used to minimise the impact of human activity.*

**By the end of this chapter, you should be able to:**

✓ describe some of the methods used to minimise the impact of human activity in the tundra
✓ give detailed examples of strategies used to minimise the impact of human activity in the tundra.

## Management strategies

There are a number of strategies used to try to reduce the amount of damage caused by human activities in tundra regions.

In Alaska, the strategies used to minimise damage caused by the pipeline include:

- Bridges were built over the pipeline so that migrating animals and people could cross.
- The pipeline was raised in some places to allow for animals to pass underneath.
- Raising the pipeline above ground also prevents the warm oil from melting the permafrost.
- Companies have high tech equipment that can be used to detect any damage to the pipeline.

**Figure 22.1**
The Trans-Alaska Pipeline and caribou

In order to protect the environments in tundra regions, countries have adopted a number of strategies. For example, Alaska has created national parks and nature reserves to protect the wildlife, ecosystems and natural landscapes. Habitat conservation programmes such as the Alaska Conservation Foundation and charities such as the WWF have all been created with the aim of protecting wildlife living in tundra regions and their habitats. Russia and Canada have developed internationally recognised programmes to protect species and their habitats, known as national Biodiversity Action Plans.

One of the biggest threats to the tundra is global warming and there are many ways of slowing down the warming of these Arctic areas. Encouragingly, most countries around the world have signed the Kyoto Protocol, which states that all countries must commit to reducing greenhouse gas emissions. Tundra regions are noticing the effects of global warming more than any other area in the world. The Kyoto Protocol will hopefully help to slow down these effects.

**Figure 22.2**
'Think Global. Act Local.'

Many countries are investing in renewable energy sources as a means of tackling the problem of global warming by reducing greenhouse gas emissions. Wind, solar, water and wave power are all being practised in countries around the world.

Strategies to tackle global warming and global climate change are not just done on an international and national level; everyone can play their part in reducing their carbon footprint. People are encouraged to 'Think Global. Act Local.' This philosophy is designed to get people to make small changes, such as recycling more, taking public transport where possible, cycling or walking where possible, using energy efficient light bulbs, insulating homes and installing solar panels. All of these actions will reduce carbon dioxide emissions.

# National 4

1. List the strategies used to minimise the damage caused by the Trans-Alaska Pipeline.
2. How is the environment protected in the tundra?
3. What can be done to slow down global warming?

# THE MANAGEMENT OF HUMAN ACTIVITIES IN THE TUNDRA

# National 5

1. Give examples of strategies used to minimise the damage caused by the Trans-Alaska Pipeline.
2. Describe, in detail, strategies used to protect the environment in the tundra.
3. Explain ways in which the effects of global warming can be reduced.

## Activities

### Activity A

Using a blank map of Alaska and an atlas, mark on the following information:

- Prudhoe Bay
- Valdez
- Anchorage
- Fairbanks
- The Brooks mountain range
- The Alaska mountain range
- Yukon River
- Kuskokwim River
- Koyukuk River
- The Trans-Alaska Pipeline.

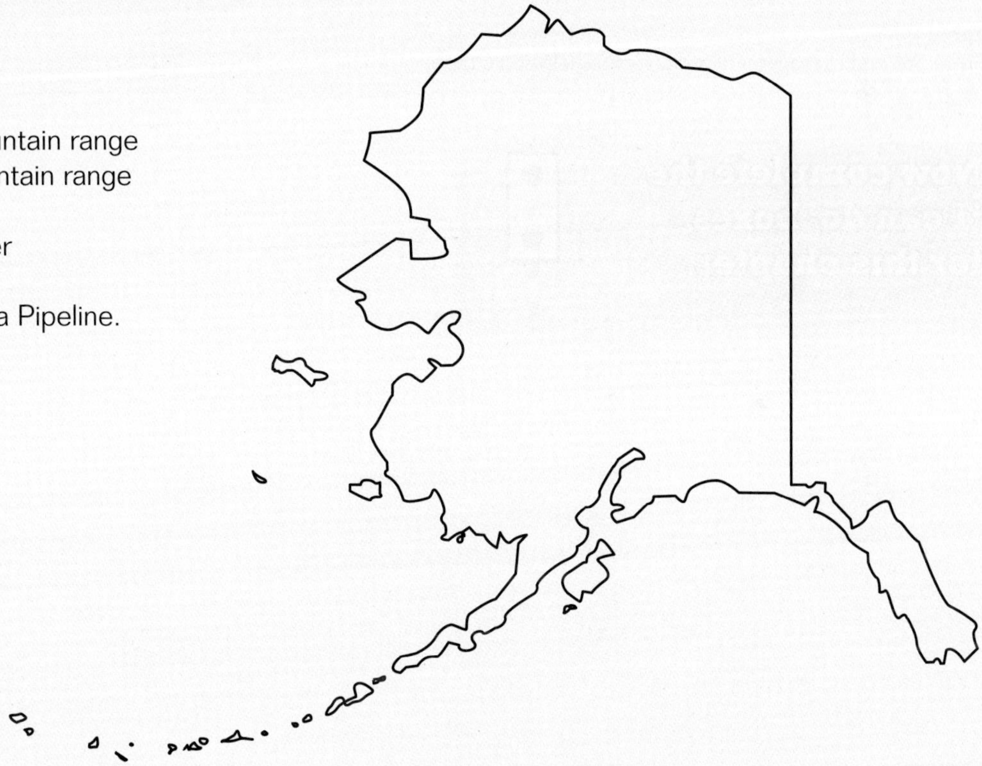

Figure 22.3
Alaska

## Activities continued...

### Activity B

Match up which 'problem' each of the management strategies would help to solve.

| Problem | Strategy |
| --- | --- |
| Melting permafrost | Building bridges for migrating animals |
| High concentrations of carbon dioxide in the atmosphere | Raising the pipeline above ground |
| High concentrations of greenhouse gases in the atmosphere | Creating national parks |
| Caribous' migration paths being affected | Kyoto Protocol |
| Destruction of habitats | Investing in renewable energy |

**Table 22.1**
Problems and management strategies to help solve them

Now complete the 'I can do' boxes for this chapter.

# Chapter 23

## The equatorial rainforest climate

This chapter looks at the equatorial rainforest climate and its ecosystem.

**By the end of this chapter, you should be able to:**

- describe the equatorial climate
- give examples of countries with an equatorial climate
- interpret a climate graph for equatorial regions
- give examples of ecosystems within equatorial regions.

## Equatorial climates

### Location

Lowland areas situated along the equator generally experience an equatorial climate. This climate region is generally found between the Tropics of Cancer and Capricorn, between 5° north and 5° south of the equator. Countries such as Brazil, Peru, Indonesia and the Democratic Republic of Congo have an equatorial climate.

**Did you know...?**
Rainforests receive over 2000mm of rainfall per year. Sometimes it can even be double that!

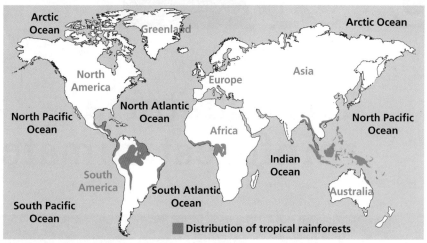

**Figure 23.1**
Location of equatorial climates and rainforests

## Climate

Tropical rainforests, such as the Amazon, grow in equatorial regions and cover approximately 6% of the Earth's surface. Equatorial climates have very specific characteristics.

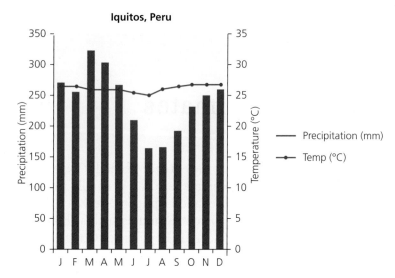

**Figure 23.2**
Climate graph of an equatorial region

Due to the latitude of the equatorial region, temperatures are high and constant throughout the year because the sun is always overhead or nearly overhead. Temperatures are generally around 27°C and do not fluctuate much. Equatorial climates tend to just have one season, therefore the weather is the same all year round. There is a lot of precipitation; most equatorial regions will receive more than 2000mm of rainfall per year. There are downpours every day as the warm, moist air is forced upwards creating low pressure, resulting in convectional thunderstorms. Evapotranspiration from the forest also assists in the build-up of cumulonimbus clouds. The hot, wet conditions make these climates very humid.

The day-to-day pattern of rainfall is the same. Between 6am and 6pm there is full daylight, temperatures are high and reach a peak around midday. At 6pm, darkness draws in and the thunderstorms begin. This is the same every day in equatorial regions.

**Figure 23.3**
Equatorial rainforest

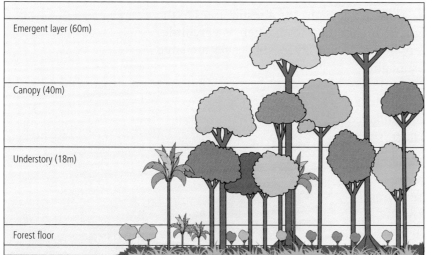

**Figure 23.4**
Layers of the rainforest

**Did you know…?**
It can take 10 minutes for a raindrop to travel from the canopy layer to the forest floor.

## Ecosystems

Despite the relatively infertile soils, equatorial regions have a huge variety of plant species growing in rainforests. There are four main layers of the rainforest.

The emergent layer is the tallest layer of the rainforest. The trees that grow here are the tallest in the rainforest, sticking out high above the canopy. Emergent trees get the most sunshine and they can grow to approximately 60m in height. Trees found in the emergent layer tend to be evergreen so they don't lose their leaves. The leaves are small and pointed and are covered in a thick waxy surface to hold water. The trees have buttress roots that help to support them and keep them from falling over as the root systems are quite shallow and the soils are quite poor. Buttress roots grow out from the trunk and spread out across the forest floor, helping to anchor the tree to the ground.

The canopy is the main layer in the rainforest and this is the layer that provides the thick cover of dense forest. Trees in the canopy can grow to approximately 40m in height and they receive a lot of sunshine. Many animals live in the canopy layer because of the shelter that the thick leafy coverage provides as well as the fruit, berries and seeds that are found here. It is estimated that approximately 90% of all animals of the rainforest are found in the canopy layer, including toucans, iguanas, sloths, spider monkeys and orangutans.

**Figure 23.5**
A buttress root

The understory is found below the canopy. It is dark in the understory as the sunlight is blocked by the canopy layer, leaving everything in shade. The understory only gets about 5% of the available daylight. As a result, trees in the understory tend not to grow very high, no higher than 18m. There is very little breeze in the understory so it feels very humid in this layer. The leaves are large and wide so that they can catch any sunlight or water that's available. Vines grow round the trees in order to reach up to the sunlight. Most reptiles living in the understory are camouflaged so that they are protected from predators, such as the red-eyed tree frog and the emerald boa.

The forest floor is the bottom layer of the rainforest and it is the darkest layer, receiving only 2% of the available sunlight. Very few plants grow on the forest floor, which is mainly made up of leaf litter. Dead vegetation, plants, leaves and roots are easily broken down and turned into organic matter (humus) which is then absorbed by tree roots. There are still insects and animals found here. Insects feed on the leaf litter, and animals such as leopards, armadillos and gorillas are found searching for food.

## The nutrient cycle

Rainforest soils are relatively poor considering the huge variety of plants that they sustain. The nutrient cycle helps to maintain the large variety of plants and trees found there. The warm, humid environment provides ideal conditions

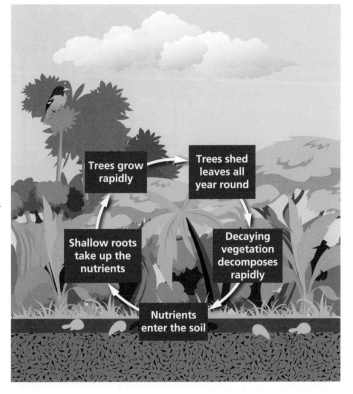

**Figure 23.6**
The nutrient cycle

for breaking down dead plants, leaves and vegetation. The decaying vegetation is turned into organic matter (humus) which is absorbed by tree roots, allowing trees and vegetation to grow. This is a continuous cycle.

When the rainforest is cut down (deforested) it interferes with the cycle as the land is left bare and the nutrients are removed (leached) by the precipitation. The top soil is eventually washed away.

## National 4

1. Where is the equatorial climate found?
2. Give three examples of countries with an equatorial climate.
3. Describe what temperatures are like in equatorial regions.
4. Describe what the precipitation is like in equatorial regions.
5. What are the four layers of the rainforest?
6. Describe three ways that plants in the rainforest have adapted to the environment.
7. Which layer has the most species living in it?
8. Which layer receives:
   a) the most sunlight
   b) the least sunlight?
9. Copy the diagram showing the nutrient cycle and label it.

## National 5

1. Describe the location of the equatorial climate, mentioning specific countries in your answer.
2. Describe, in detail, the equatorial climate.
3. For each layer of the rainforest, write down its main characteristics.
4. Explain ways in which vegetation has adapted to the environment.
5. Make a copy of the nutrient cycle and describe it.

## Activities

### Activity A

Using the information in the table below, draw a climate graph for an equatorial region.

|  | Jan | Feb | Mar | Apr | May | Jun | Jul | Aug | Sep | Oct | Nov | Dec |
| --- | --- | --- | --- | --- | --- | --- | --- | --- | --- | --- | --- | --- |
| Temperature (°C) | 26 | 26 | 27 | 27 | 26 | 28 | 28 | 27 | 27 | 26 | 27 | 27 |
| Precipitation (mm) | 550 | 525 | 550 | 620 | 650 | 660 | 540 | 550 | 540 | 555 | 540 | 550 |

**Table 23.1**
Climate of an equatorial region

## Activities continued

### Activity B

Photocopy the figure below. Cut out each of the dominoes.

With a partner, split the dominoes equally between you. Lay them out so that you can see what is on each of your dominoes. One person starts and then you take it in turns to lay a domino down. You must match the correct term with the right definition.

| START | EMERGENTS | RAINFORESTS | EVAPOTRANSPIRATION |
|---|---|---|---|
| Thick, dense forest layer that blocks out sunlight to lower layers | The process where moisture is lost from vegetation and turns to water vapour in the atmosphere | Trees that don't lose their leaves | Heavy downpours that occur every day in the rainforest |
| **EQUATORIAL CLIMATE** | **THUNDERSTORMS** | **CUMULONIMBUS** | **EVERGREEN** |
| Lines of latitude found 23.5° north and south of the equator | The name given to the tallest trees in the rainforest | The name given to organic matter found in the top soil | The name given to large tree roots |
| **BUTTRESS ROOTS** | **CANOPY** | **HOT, WET, HUMID** | **AMAZON** |
| This layer receives about 5% of daylight and has mainly small trees | Rain clouds | Dense forest, rich in biodiversity, found between the tropics | The type of weather that is found in rainforests |
| **BRAZIL** | **HUMUS** | **TROPICS** | **UNDERSTORY** |
| The removal of nutrients from the top soil by precipitation | The type of climate that allows tropical rainforests to grow | This layer is dark and damp and contains decaying plants and roots | The cutting down of trees |
| **FOREST FLOOR** | **LEACHING** | **VINES** | **DEFORESTATION** |
| The name of a tropical rainforest found in South America | These plants grow around trees in order to reach sunlight | A country that has tropical rainforests | FINISH |

**Now complete the 'I can do' boxes for this chapter.**

# Chapter 24

## Uses of the equatorial rainforest

*This chapter looks at the uses made of the equatorial rainforest.*

**By the end of this chapter, you should be able to:**

✓ describe the traditional uses of the equatorial rainforests
✓ describe recent activities in the equatorial rainforests.

**Did you know…?** 25% of all modern medicines come from the rainforest.

## How the rainforest is used

Rainforests used to cover approximately 14% of the Earth's surface; today this has reduced to only 6%. Rainforests are not just home to plants and animals, there are still some people that call the rainforest home.

People living in rainforests, such as the Amazon, live very traditional lifestyles. Native people such as the Yanomami and Pygmies have lived in rainforests for thousands of years. There is an abundance of plant and animal life there that makes living in the rainforest possible. The indigenous (native) people have many practices but none of them damage the rainforest. One of their main practices has traditionally been shifting cultivation. This is a specific farming method which is still popular in rainforest regions.

## Shifting cultivation

Shifting cultivation is a way of farming. Soils are relatively poor in rainforests so this practice has developed in order to get as much from the land as possible. Indigenous people clear a small area of land by a process known as 'slash and burn'. This is where vegetation is cut back using machetes and axes. Large trees are not usually cut down but are kept and used for their berries and their leaves. The vegetation is then burnt. The ash that is produced from the burning of the vegetation is used as a natural fertiliser that helps crops to grow in the poor soils. Crops such as sweet potato, plantain and manioc are then grown. The first crop yield is usually quite productive as the soils have been improved by the ash, however as more crops are grown the quality of the soil deteriorates so people only stay at a location for a few years before moving on. When the tribal group moves on, they find a new location and the same sequence of events takes place. The previous sites are left to regenerate, so that when the group returns, the vegetation has grown back.

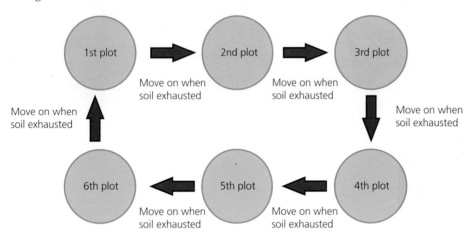

**Figure 24.1**
The process of shifting cultivation

The practice of shifting cultivation does not damage the rainforest. These practices have been going on for thousands of years and the rainforests have always regenerated. This type of farming is sustainable.

## Recent activities in the equatorial rainforests

Rainforests have an abundance of resources and this made them vulnerable to outside influences. Almost half of the world's rainforests have been cleared in the last few decades and they continue to be under threat.

Every year 81 million hectares of rainforest are cut down, the equivalent of 36 football pitches per minute. They are cut down to make room for a variety of human activities.

The biggest reason for deforestation is for farming. It is believed that 80% of deforestation is for this purpose. Due to the poor soils in rainforests, arable farming has to be done on a huge scale to be profitable – so large areas of rainforests are cleared. However, the largest driver behind deforestation is cattle ranching. The Amazon rainforest region in Brazil is the largest exporter of cattle in the world.

# USES OF THE EQUATORIAL RAINFOREST

**Figure 24.2**
Cattle ranching in the Amazon

Mining is another driving force behind the deforestation. Gold, copper and diamonds are found in rainforest regions. The value of such resources means that mining them is a profitable industry. Vast areas of rainforest are cleared to allow mining to take place. Areas are not just cleared for the mines themselves but also for road networks to be built through the rainforest to allow for transportation of these precious metals.

The timber itself creates a profitable industry. Logging is big business as many of the trees (such as teak and mahogany) fetch high prices. Trees are felled and transported by road, rail and river to ports for export. Some are also needed in nearby settlements as building materials for new homes or to be made into furniture.

The world's population is increasing and that brings pressure to build more houses and settlements. Large areas of rainforests are cleared to build new settlements.

While people may benefit from the deforestation of rainforests as they may be employed in the industries or live in the new houses and settlements, overall this large-scale deforestation is more harmful than traditional shifting cultivation. The problems it causes are described in the next chapter.

## National 4

1. What percentage of the Earth do rainforests cover today?
2. Give examples of native tribes living in the rainforest.
3. What is meant by the term 'slash and burn'?
4. Why are large trees not cut down?
5. Why is the vegetation burnt?
6. Give examples of crops that are grown in the rainforest clearings.
7. Why do people only stay at a clearing for a few years?
8. Copy Figure 24.1 showing the process of shifting cultivation and label it.
9. List the main reasons why rainforests are being cut down.

## National 5

1. Using a diagram, describe, in detail, the shifting cultivation method of farming.
2. Why is shifting cultivation considered sustainable?
3. Explain why large-scale deforestation is now taking place.

USES OF THE EQUATORIAL RAINFOREST

## Activities

### Activity A

Create a series of pictures showing the process of shifting cultivation. Your pictures should describe the process from start to finish.

### Activity B

Match each of photos A–E with the correct reason for deforestation listed below.

1. Logging
2. Cattle ranching
3. Settlement
4. Mining
5. Communications

A

B

C

D

E

Now complete the 'I can do' boxes for this chapter.

# Chapter 25

*This chapter looks at the effects of deforestation on the people and environment of the Amazon rainforest.*

# The Amazon rainforest: effects of deforestation

**By the end of this chapter, you should be able to:**

- ✓ give examples of how deforestation affects people
- ✓ describe the effects of deforestation on the environment.

## The extent of deforestation

Deforestation is simply the cutting down of trees. For centuries, rainforests were barely touched by humans and their extent remained the same; it is only in the last 50 years that they have been subject to rapid destruction.

Rainforests used to cover approximately 14% of the Earth's surface; today it is only 6% and it is decreasing at an alarming rate. One example of this is in the Amazon rainforest in South America. As we have seen, the reasons for this destruction are, chiefly, to make way for large-scale cattle ranches, new settlements and industry. The effects that these wide-scale clearances have on the people and the environment are wide-ranging.

**Did you know…?**
An area of rainforest the size of 36 football pitches disappears each minute!

Figure 25.1
Deforestation

## The effects of deforestation on people

Since 1978 almost 800,000 square kilometres of rainforest have been destroyed in the Amazon. Countries such as Brazil, Bolivia and Peru have seen huge areas cleared to make way for new activities. Indigenous people living in the Amazon, such as the Kayapo, lived their lives completely unaware of anything outside of the forest. For centuries, there have been indigenous people living inside the rainforest following very traditional lifestyles. Since the rapid deforestation of these environments, indigenous people have 'lost' their old ways of life and have had to adapt their lifestyles. With the destruction of the rainforest goes the culture, skills and knowledge that native people have built up over many centuries. As more and more land is lost, indigenous people have had to alter their practice of shifting cultivation, visiting fewer sites and returning to old sites more quickly than before. This means the land does not support as many people and some are forced to move away to nearby towns. Very few indigenous people living in rainforests remain unaware of 'outsiders'; most native people will now have come across people on the outside. This can be dangerous for the indigenous tribes as they have no resistance to diseases that outsiders pass on to them, resulting in serious illness for them.

Figure 25.2
Indigenous women in the rainforest

# The effects of deforestation on the environment

The Amazon rainforest covers approximately 670 million hectares across nine different countries. Tropical rainforests are nicknamed the 'lungs of our planet' as they absorb carbon dioxide and breathe out oxygen. This is essential for the health of our climate. When rainforests are destroyed, there are fewer trees to absorb the $CO_2$ and any stored carbon that the tree contains is released into the atmosphere, increasing carbon dioxide concentrations in the atmosphere. If rainforests are burned down, this process further emits large concentrations of carbon dioxide into the atmosphere. It is thought that deforestation is one of the major contributory factors to global climate change.

As more and more rainforests are cleared, the land where they once were is turning to desert. The roots of the vegetation in rainforest regions such as the Amazon bind the soil together. When the vegetation is destroyed, the soil becomes loose and the top layers are washed away; this is more commonly referred to as soil erosion. The top layers of the soil contain the most nutrients so when these are lost as the soil is washed or blown away, the land becomes infertile. If the soil is washed into nearby rivers, it can cause the rivers to silt up and can lead to wide-scale flooding. The deforestation also interferes with the local water and nutrient cycles. As less water vapour is returned to the atmosphere through the vegetation, less convectional rain falls, so the area becomes drier. The land eventually becomes more arid.

Loss of species is one of the biggest impacts that deforestation has on the environment. Approximately 70% of all plants and animals live in rainforests. As the land is cleared, so too are the animal habitats. This can lead to species endangerment and eventually extinction. In the Amazon, 38 species have become extinct in the last 30 years as their hunting and breeding grounds have been reduced. Species such as the white-cheeked spider monkey and the bare-faced tamarin are among those species living in the Amazon that are under threat.

Many of our western medicines come from rainforest plants. The loss of these plant species also means the loss of potential cures and treatments for many of the world's diseases.

Once trees have been destroyed it takes years for them to grow back. Some types of trees take a very long time to grow and once they have been cut down they take a long time to grow back.

# National 4

1. Make a list of the ways that deforestation affects people.
2. In what way does deforestation lead to an increase in carbon dioxide levels?
3. In what way does deforestation cause soil erosion?
4. Why should less rain fall after deforestation?

# National 5

1. Describe, in detail, the impact of deforestation on the indigenous people of the Amazon region.
2. In what ways does deforestation affect global climate change?
3. Complete the following:
    a) Make a list of the ways that deforestation affects the environment.
    b) Choose two of these and explain them in detail.

## Activity

Write a report on the effects of deforestation on people and the environment.

a) Begin by mentioning where tropical rainforests are found, what benefits they bring to our planet and what deforestation is.
b) Then, describe the negative effects that cutting down rainforests has on the people who live in them.
c) Go on to describe the negative effects of deforestation on the environment, commenting on how deforestation is linked to climate change.
d) Mention specific named examples where possible.
e) Include diagrams and pictures in your report.

Now complete the 'I can do' boxes for this chapter.

# Chapter 26

## The management of human activities in equatorial rainforests

*This chapter looks at ways to minimise the damage caused by deforestation in equatorial regions.*

**By the end of this chapter, you should be able to:**

- give examples of management strategies used in the Amazon region
- explain how these strategies are helping to minimise the damage caused by deforestation.

## Management strategies

Some believe that if current rates of deforestation were to continue, the world's rainforests would be completely gone in 100 years' time. While rates of deforestation remain high, there are a number of projects and strategies that have been designed to minimise the effects of deforestation.

Sustainable forestry is practised in a number of countries, in particular Malaysia, which has a large area of tropical rainforest. Sustainable forestry means that trees can be cut down but they have to meet certain criteria and the deforestation must be manageable and not negatively affect the forest in the long-term. When logging companies select trees to cut down, they must make sure that they:

- don't cut down trees that are under a minimum height or circumference
- restrict felling so that they ensure enough trees per hectare are remaining
- restrict the use of heavy machinery such as bulldozers as these can cause further damage to the rainforest

- use roads/tracks that are at least one kilometre apart
- use alternative methods of transportation. Huge lorries are needed to transport the timber from the rainforests. Building roads large enough to accommodate these vehicles is equally destructive. Heli-logging is a way of removing trees from the rainforest without using large vehicles. However, this is very expensive.

Other strategies that have been adopted include:

- Creating national parks, as this can protect the forests and restrict or prohibit deforestation within the boundaries.
- Afforestation schemes whereby, for every tree that is cut down, one is planted. This will help to maintain the forest in the future.
- Educating everyone involved in the cutting down or exploiting of the rainforests. Making these people aware of the damage that deforestation causes and educating them on how they can minimise the impact that they have.
- Monitoring of the forest by satellite, which allows governments/charities to observe the activities that are taking place there.

**Figure 26.1**
Heli-logging, shown here in a forest environment, is also an option for log transportation in equatorial rainforests

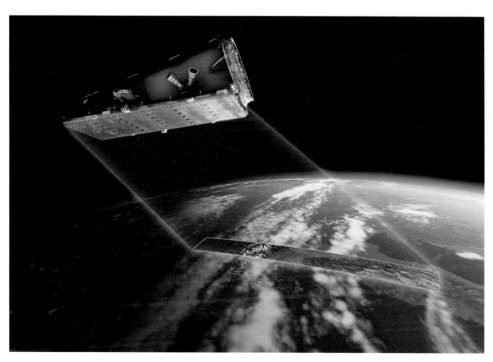

**Figure 26.2**
Monitoring deforestation

## National 4

1. What is sustainable forestry?
2. What factors must logging companies consider when selecting trees?
3. What other strategies have been used to protect rainforests?
4. In your opinion, what are the best ways that forests can be protected? Give reasons for your answer.

## National 5

1. Describe, in detail, what is meant by the term 'sustainable forestry'.
2. What factors must logging companies consider when selecting trees?
3. Give examples of ways that sustainable forestry works.
4. What other strategies have been used to protect rainforests?
5. In your opinion, what are the best ways that forests can be protected? Give reasons for your answer.

### Activity

Design a leaflet that can be given to logging companies that practise illegally. Highlight to them what measures should be taken when selecting trees for cutting down and give examples of how they can minimise damage caused to the rainforest.

Now complete the 'I can do' boxes for this chapter.

# N5 Examination questions

Read the information and the two SQA-style questions below, and answer Tasks A–H.

## Question 1

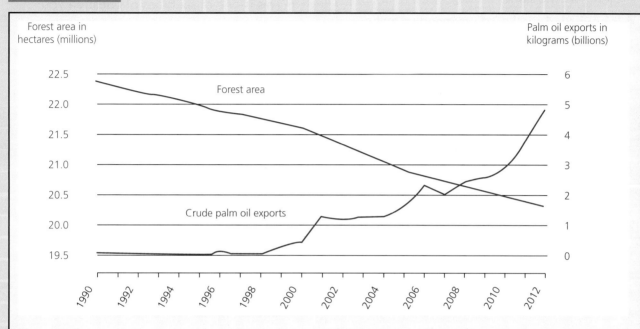

**Figure 26.3**
Diagram Q1 – Deforestation and palm oil exports in Malaysia (1990–2012)

(a) Study Diagram Q1. Describe the changes in exports of palm oil and the forest area in Malaysia between 1990 and 2012. (4)

(b) For an area you have studied in the equatorial or tundra region, give advantages and disadvantages brought about by recent human activities. (6)

### Advice for Question 1(a)

You should have spotted four important points.

1 It is a *describe* question.

2 It has a diagram – Diagram Q1. Your task is to interpret both lines on the graph.

### Advice for Question 1(b)

You should have spotted four important points.

1 It is a *give advantages and disadvantages* question.

2 It has no diagram.

3 You must make a choice.

4 It is worth **6** marks.

## Advice for Question 1(a)

3 It is a *describe the changes* question. You do not get marks for giving the amounts but for saying whether they increased, decreased or stayed the same.

4 It is worth **4** marks, so you need to make 4 factual points or at least 2 developed points. To make a developed point, you would need to quote figures and years. For example, *Crude palm oil exports increased quickly from 1999, the fastest increase being between 2000 and 2001 when they went up from 0.4 to 1.3 billion kg.* (2)

**TASK A**: Read the advice and then answer Question 1(a).

## Advice for Question 1(b)

Let's take these points separately.

1 It is a *give advantages and disadvantages* question.
   - This is one of six types of question you can be asked. Sometimes the question will only ask you for advantages or for disadvantages.
   - You must show your understanding by explaining a few benefits and problems caused by human activities. This means you must make it clear how each activity has affected the region.

2 It has no diagram.
   - You need to answer this question from your knowledge of the coursework.

3 You must make a choice.
   - It is easy to miss, but the question tells you to write about the equatorial **or** the tundra region.

4 It is worth **6** marks.
   - A *give advantages and disadvantages* question is usually worth 4–6 marks, and you are expected to answer in sentences.
   - There is 1 mark for a briefly explained point but you can earn 2 marks by developing the point. For example, *One advantage of mining in the tundra is that it brings many quite well-paid jobs* (1) *so people have money to spend which means there are more jobs in shops, entertainments and restaurants.* (1)

**TASK B**: Read the advice and then answer Question 1(b).

# Question 2

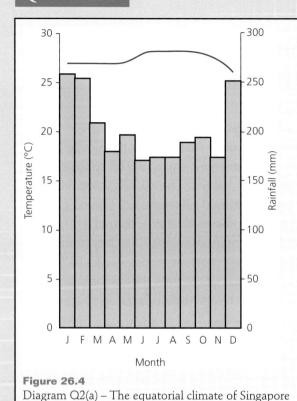

**Figure 26.4**
Diagram Q2(a) – The equatorial climate of Singapore

1. **Afforestation** – plant more trees to replace those cut down
2. **Education** – persuade companies not to deforest and educate the local people about its harmful effects
3. **A government ban** on all logging and road-building
4. **Set up national parks** where the rainforest is protected
5. **Better policing** – use more people and technology, such as satellites, to catch illegal loggers and punish them

**Figure 26.5**
Diagram Q2(b) – Methods of controlling deforestation in equatorial regions

(a) Study Diagram Q2(a). Describe, in detail, the main features of the equatorial climate of Singapore. (4)

(b) Choose two strategies from Diagram Q2(b). For each, describe its advantages and disadvantages in controlling deforestation in the equatorial regions. (6)

## How good is this answer to Question 2(a)?

In January the temperature is 27°C and the rainfall just over 250mm. By March the temperatures are still 27°C but the rainfall is down to 200mm. By July the temperature has risen to 28°C but the rainfall has dropped further to 180mm. In October the temperature is still 28°C but the rainfall has almost climbed back to 200mm.

**TASK C**: Read the answer above and give it a mark. Explain the number of marks you have given.

## How good is this answer to Question 2(b)?

If you made the forest a national park then it could not be destroyed which would be good for the environment but it would still need people to check it so it would still cost money and you can't take out valuable things so people don't have as much money and nor does the government so it might be better to let them cut down the trees and then plant new ones afterwards or plant trees and crops so people can carry on farming as well.

**TASK D**: Read the answer above and give it a mark. Explain the number of marks you have given.

## Question 1 on page 143

### This is a good answer to Question 1(a)

The forest area decreased between 1990 and 2012 from 22.4 to 20.3 million ha. (1) It was a steady decrease but slightly faster between 2001 and 2005 when it decreased by 0.8 million hectares. (2)

There were almost no palm oil exports between 1990 and 1998. They then increased quickly from 1999, the fastest increase being between 2000 and 2001 when they went up from 0.4 to 1.3 billion kg. (2) Between 2006 and 2007 there was the only decrease. (1)

### Why this is a good answer

The candidate has written separately about the 2 graph lines and has focussed on changes. They have not only quoted figures and years but have also, in the second sentence, subtracted numbers to give a precise change. This makes for a developed point.

**TASK E**: Can you improve even further on the above answer by giving more subtractions and/or additions?

### This is a good answer to Question 1(b)

The tundra
One human activity has been mining, for example for oil in Alaska. This has brought many disadvantages, especially oil spillages such as the Exxon Valdez. The oil kills water life and the birds that feed off them. (2) The oil gets into plants and can kill them, so there is less food for the animals to eat. (1) They have had to build a huge pipeline across Alaska to get the oil out, but this upsets the caribou which have had to change their migration routes. (1)

On the other hand, a big advantage of mining is that it brings many quite well-paid jobs so people have money to spend which means there are more jobs in shops, entertainments and restaurants. (2) But even this has a disadvantage because the next generation are forgetting the old ways of making a living and the culture of the area. (1)

### Why this is a good answer

The candidate has given their choice at the start. The answer also includes both advantages and disadvantages. These are clearly explained in detail which makes many of them developed points. The candidate has also referred to specific places (Alaska) and mentioned other specific names (Exxon Valdez, caribou).

**TASK F**: This is not a perfect answer, chiefly because it just gives the advantages and disadvantages of one activity. Because of this, it might not earn full marks. Add the good and bad effects of two more recent human activities to this answer to make sure it earns full marks.

# N5 EXAMINATION QUESTIONS

147

## Question 2 on page 145

### Advice for Question 2(a)

You should have spotted two important points.

1. It is a *describe* question and it has a diagram – Diagram Q2 (a). You are expected to interpret the diagram – the climate graph. You should state the main patterns of the temperature and rainfall. It is common to give the highest and lowest figures and even the range.
2. It is worth **4** marks, so you must make 4 valid points or 2 developed points. A valid point would be to state: *The highest rainfall is in January, just over 250mm.* (1) To make a developed point you would need to describe the range of monthly rainfall figures.

### How good was the answer on p.145?

**(Mark: possibly 3 out of 4)**

The answer accurately interprets the graph, correctly giving the temperature and rainfall in four months. This is sufficient to earn most of the marks, but it is not the best answer. Someone wanting to know what this climate was like would prefer a summary, not a month-by-month list of figures. So the candidate should summarise the main points of the temperature and rainfall. For temperature for example, *Temperatures hardly change at all, only from 26°C in December to 28°C from June to October.* (1)

**TASK G**: Read the comments above and then write an improved answer to Question 2(a).

### Advice for Question 2(b)

You should have spotted three important points.

1. It is a *give advantages and disadvantages* question. So you must mention a few benefits and problems of the different strategies.
2. You must make a choice. In this case, you must choose two strategies from the list; you cannot bring in other strategies that you have learned.
3. It is worth **6** marks, so you need to make 6 separate points or, if possible, 3 more detailed points.

### How good was the answer on p.145?

**(Mark: 3 out of 6)**

This is an awkward answer to mark. The answer does give advantages and disadvantages of two strategies, which is what the question asks. It gives 6 points but the last point relates to agro-forestry, which was not one of the choices. The candidate has rambled, the whole answer being one sentence and the points made being mostly simple ones. If they had made each point into a separate sentence, it would have made it possible to develop the point and earn more marks. As it stands, only 3 of those points deserved a mark and none were developed.

**TASK H**: Read the comments above and then write an improved answer to Question 2(b). Make each of the first 5 points into a separate sentence and try to develop each point.

# Chapter 27

## Health in developing countries

This chapter looks at how healthy people are in developing countries.

**By the end of this chapter, you should be able to:**

- ✓ describe the main reasons for ill-health in developing countries
- ✓ explain the effects of ill-health on the people and their countries
- ✓ describe some solutions to ill-health in developing countries.

## Health comparisons between developing and developed countries

|  | 1950 | 1970 | 1990 | 2010 |
|---|---|---|---|---|
| Developing countries | 41 | 52 | 61 | 67 |
| Developed countries | 66 | 70 | 74 | 78 |

**Table 27.1**
Changes in life expectancy (1950–2010)

Table 27.1 shows one of the starkest differences between the worlds of the rich and poor. Health is much better in the rich world and people live far longer. If you are born and live your life in Japan you can expect to live 36 years longer than if you live in the African country of Chad. Of all the inequalities in the world, this is surely one of the worst.

**Did you know…?**
Six countries in Africa had lower life expectancies in 2010 than in 1970.

# HEALTH IN DEVELOPING COUNTRIES

The quality of people's health is used to indicate the level of development of a country; if many people are suffering ill-health, this indicates a low level of development, and vice versa. Ill-health makes people very weak and may leave them with a permanent disability, which can result in them quickly becoming trapped in a vicious cycle of disease (see Figure 27.1). In developing countries most people suffer from at least one disease. Not only does this reduce the quality of their lives but it also seriously reduces the economic development of the whole country. This is because:

- people who are unwell cannot work and therefore don't produce wealth
- people who are unwell need other people to look after them – people who otherwise would be working themselves.

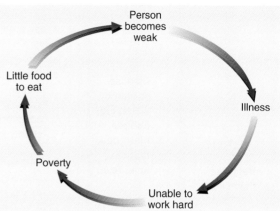

**Figure 27.1**
Vicious cycle of disease

## Causes of ill-health

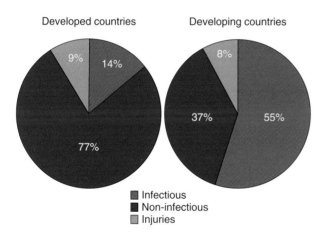

**Figure 27.2**
Causes of death in developed and developing countries

**Did you know....?**
Six infectious diseases – pneumonia, tuberculosis, diarrhoea, malaria, measles and HIV/AIDS – cause 90% of all deaths in developing countries.

Diseases can be divided into those that are infectious (where one person infects another) and non-infectious (those which cannot be 'caught' from someone else). In developing countries infectious diseases are far more common and account for most causes of death (see Figure 27.2). Figure 27.3 shows the causes of the most common diseases in developing countries.

### Factors in ill-health in developing countries

Drinking **polluted water**.
Lack of proper **sewage disposal** (so the sewage mixes with drinking water).
**Climate** – warmer climates attract more flies and mosquitoes which spread disease.
**Poor health care** – to prevent, treat and cure people.
**Poor diet** – not eating enough food or not eating a balanced diet.
**Poverty** – people may not have enough money to eat properly or buy medicines and the whole country may not have enough money for proper sewers, water pipes and hospitals.

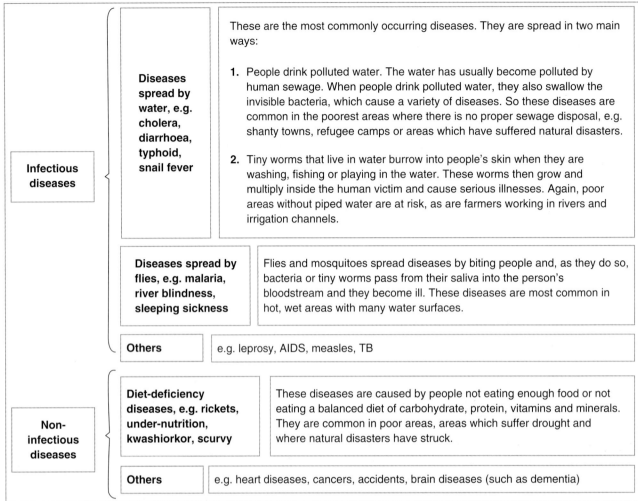

**Figure 27.3**
Diseases of the developing world

# Effects of ill-health

Diseases that are common in the developing world affect people in a variety of ways:

- **Some are killer diseases**, for example AIDS and malaria.
- **Some make people very weak** and lethargic but are not directly responsible for their deaths, for example snail fever and kwashiorkor. If people are very weak, they quickly become trapped in a vicious cycle of disease.
- **Some leave their victims with a permanent disability** or injury, for example river blindness and rickets. Not only does this make it extremely difficult for those affected to find work, but it also requires someone in their community to look after them.
- **Most diseases reduce people's life expectancy**, either directly or indirectly.
- **All diseases reduce the rate of development of the country** and the wealth of the people.

# HEALTH IN DEVELOPING COUNTRIES

## Improving health

Most of the diseases that affect people in the developing world can be prevented or cured. So what are the best strategies for improving health in developing countries?

- **Produce more food** – this reduces diet-deficiency diseases and makes people stronger and better able to fight off diseases.
- **Improve health facilities** – so that everyone, including those living in rural areas, has access to a health centre, trained medical aid and medicines.
- **Provide clean water** – by improving water supplies and sewage disposal.
- **Provide health education** – so people know what causes disease and the simple ways it can be prevented.

## National 4

1. Look at Table 27.1.
   (a) What was the difference in life expectancy between developed and developing countries in 2010?
   (b) Is the difference becoming greater or smaller?
2. Which are more common in developing countries – infectious or non-infectious diseases?
3. (a) Name two infectious diseases.
   (b) Name two non-infectious diseases.
4. Describe the vicious cycle of disease shown in Figure 27.1.
5. (a) Explain how polluted water spreads disease.
   (b) Explain how flies spread disease.
6. What causes diet-deficiency diseases?
7. Explain why poverty is an important factor in health.
8. Explain two ways of improving health in developing countries.

## National 5

1. Describe what is shown in Table 27.1.
2. Look at Figure 27.2.
   (a) Describe the differences in causes of death between developed and developing countries.
   (b) Explain fully how polluted water spreads disease.
   (c) Explain fully how flies spread disease.
3. What are the main causes of non-infectious diseases in developing countries?
4. 'Poverty causes ill-health and ill-health causes poverty.' Explain what this statement means.
5. Which diseases would be reduced by providing clean water to a community?

## Activities

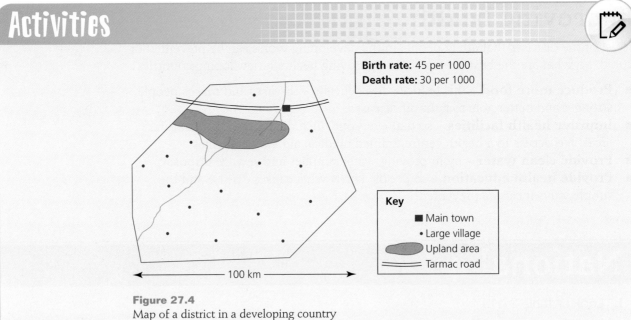

**Figure 27.4**
Map of a district in a developing country

### Activity A

The government of a developing country has made health a priority. It has set aside an extra amount of money for its very poorest farming district, which has the lowest life expectancy in the country (see the above figure). The government knows there are three big health problems but it is unsure how to tackle them. The problems and possible solutions are shown below.

1. Diseases spread by polluted water. *Either:* install sewers so that sewage does not mix with the water supply, *or:* provide free treatment to people suffering from these diseases.
2. Poor diet. *Either:* install small reservoirs (tanks) which allow people in each village to store water, *or:* give a free meal daily to everyone.
3. Lack of health facilities. *Either:* build a small health centre in every large village (with a nurse who can give health advice and provide cheap medicines), *or:* build one large hospital in the biggest town (with doctors, better equipment, surgery and maternity facilities).

For each of the three problems, decide which of the two solutions you think is best. Justify your answer.

### Activity B

Unfortunately the government has less money than it expected and can only afford to tackle one of the above problems. Which one should it tackle, and why? (Think what health problems each would solve and the knock-on effects of solving these.)

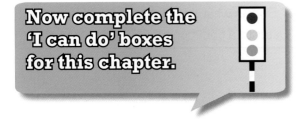

Now complete the 'I can do' boxes for this chapter.

# Chapter 28

## Health in developed countries

*This chapter looks at how healthy people are in developed countries.*

**By the end of this chapter, you should be able to:**

- ✓ explain why developed countries have less disease than developing countries
- ✓ describe the areas within developed countries which have the worst health
- ✓ describe the main factors affecting health in developed countries.

## Health in developed countries

As mentioned in Chapter 27, people in developed countries can expect to live on average 11 years longer than people in developing countries. There are three main reasons for this:

1. Clean environmental conditions
2. High-quality health facilities
3. Improved health education

Table 28.1 compares various indicators of health between developed countries and developing countries.

## Clean environmental conditions

People in developed countries live in a cleaner, healthier environment. Our water is purified before it reaches our taps. Sewage is taken away by pipes and treated before being emptied into rivers and seas. Rubbish is collected regularly and disposed of properly. Under these sanitary conditions, infectious diseases are much less likely to spread.

|  | Gross national income per person ($) | % people with improved sanitation | Doctors per 10,000 people | Calories per person per day |
| --- | --- | --- | --- | --- |
| Norway | 88,890 | 100 | 42 | 3460 |
| Australia | 49,130 | 100 | 30 | 3190 |
| Canada | 45,560 | 100 | 20 | 3530 |
| Japan | 44,900 | 100 | 21 | 2810 |
| Ireland | 39,930 | 100 | 32 | 3530 |
| Pakistan | 1120 | 48 | 8 | 2250 |
| Kenya | 820 | 32 | 2 | 2060 |
| Haiti | 700 | 17 | 3 | 1850 |
| Ethiopia | 370 | 21 | 0.2 | 1950 |
| DR Congo | 190 | 18 | 1 | 1590 |

**Table 28.1**
Effects of a country's income on factors affecting health

## High-quality health facilities

Wealthier countries can afford to spend a lot of money on medical care. A full range of health care is available, including surgical operations, organ transplants and antenatal care, provided by a range of health personnel including doctors, nurses, midwives and physiotherapists. Children are routinely inoculated against diphtheria, tetanus and polio when they are a few months old; against measles, mumps and rubella when one year old; and against tuberculosis when about 13 years old. To ensure that people are not affected by the vicious cycle of disease, sickness benefit is paid to people who are unable to work due to illness.

## Improved health education

People in the developed world are much more aware of the causes of disease and how they can be prevented. We know the importance of a healthy diet, regular exercise and safe sex. It is much easier to inform people of health matters in richer countries. Countless radio programmes and TV chat shows discuss topical health issues. Health services regularly run campaigns and they get their message across through advertisements in newspapers, on television, in schools and on roadside billboards.

# Ill-health in developed countries

Wealthier countries have been very successful in reducing infectious diseases, such as whooping cough, diphtheria, scarlet fever and tuberculosis. We have learned how to prevent and cure these diseases and they now account for relatively few deaths. Instead, it is the non-infectious diseases such as heart disease, type 2 diabetes and cancer that are more serious in developed countries. These are more difficult to prevent and cure, mostly because we do not fully understand their causes.

# Health in developed countries

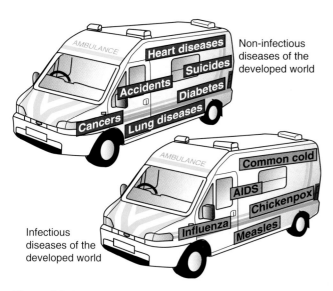

**Figure 28.1**
Examples of infectious and non-infectious diseases of the developed world

Five factors are particularly important in explaining the distribution of non-infectious diseases in developed countries:

1. Pollution
2. Social habits
3. Poor diet
4. Stress
5. Access to health facilities.

## Pollution

Air pollution from vehicles, factories and power stations is more serious in developed countries. Breathing in polluted air can cause lung disease and some forms of cancer. These diseases are more common in:

- urban areas rather than rural areas
- areas with a high concentration of heavy industry
- countries where environmental laws are less strict.

## Social habits

Social habits such as smoking and drug and alcohol abuse increase our chances of dying from a variety of diseases, such as lung cancer and liver disease. Almost 3000 people in Scotland died from alcohol-related causes in 2012. Research also shows that smokers die ten years earlier on average than non-smokers. These habits are expensive and so are more common in richer countries but, within developed countries, the highest rates of smoking and alcohol-related illnesses are found in more deprived areas.

## Poor diet

Unlike in the developing world where people suffer from a lack of food, in developed countries people suffer ill-health from eating too much, especially fatty foods. This puts us at risk of heart disease and cancer. Research has shown that 30% of children and 65% of adults in Scotland are overweight and that Scottish children are more inactive, unfit and overweight than ever before. As a result, their life expectancy may be less than that of their parents. Cheaper foods often contain more fat so, within developed countries, ill-health due to a poor diet is also more common in poorer areas.

## Stress

The faster, more hectic pace of life in developed countries affects our health. Stress is linked to heart disease, brain diseases (such as depression), even accidents and suicides. Generally, stress levels are greater for those living in cities than in the countryside.

**Figure 28.2**
Poor diet is often a cause of disease in developed countries

## Access to health facilities

Within the developed world there are significant differences in health facilities available to people in different countries and in different parts of a country. For example, the USA spends four times as much per person on health than Spain, and 100 times more than Poland. But in many developed countries, including the USA, health care has to be paid for and, as a result, life expectancy is lower in poorer areas. A report by the NHS in 2011 found that life expectancy for men in Lenzie, just outside Glasgow, was 82 years while in the Calton area of Glasgow it was only 54 years.

# National 4

1. Name four non-infectious diseases more common in developed than developing countries.
2. Why are infectious diseases less likely to spread in developed countries?
3. Describe how developed countries protect their children from catching diseases.
4. In what ways are people in developed countries educated about health?
5. Five main factors that cause ill-health in developed countries are pollution, stress, social habits, poor diet and access to health facilities. Which of these affect people in:
   (a) urban areas more than rural areas?
   (b) poorer areas more than richer areas?

# National 5

1. (a) Name three infectious and three non-infectious diseases common in developed countries.
   (b) Which account for more deaths – infectious or non-infectious diseases?
2. Why are infectious diseases less likely to spread in developed countries?
3. Explain how we are able to prevent diseases in the developed world more easily than in the developing world.
4. Life expectancy in cities in developed countries is lower than in the countryside. Explain why.
5. Life expectancy is lower in poorer areas of developed countries. Explain why.

HEALTH IN DEVELOPED COUNTRIES

## Activities

### Activity A

Pollution, social habits, poor diet and stress are four main causes of ill-health and death in rich countries.

(a) Which do you think is the biggest cause? Why?
(b) Which do you think is the easiest to reduce? How would you do it?

### Activity B

A

B

Health messages encouraging the use of seat belts (the Clunk-Click seat belt campaign) and to use sun protection to prevent overexposure to the sun (the Slip! Slop! Slap! sun protection campaign)

Give examples of health messages you can remember.

(a) How were they advertised (e.g. on TV, on posters)?
(b) How effective do you think they were?

Now complete the 'I can do' boxes for this chapter.

# Chapter 29

## Malaria – its cause and transmission

This chapter looks at malaria, a common disease in developing countries.

**By the end of this chapter, you should be able to:**

✓ describe the distribution of malaria around the world
✓ explain the causes of malaria
✓ describe how malaria is transmitted.

Infectious diseases are more common in poorer countries and non-infectious diseases are more common in rich countries. To show the effects of these diseases, the most common ones affecting both developing and developed countries will now be studied in more detail. The first disease we will look at affects 400 million people and kills 1.2 million every year, half of them children. This disease is called **malaria**.

## Cause and method of transmission of malaria

**Malaria is caused by a tiny parasite** that finds its way into a person's bloodstream. After a few days, the infected person suffers headaches and stomach pains, followed by fevers of high temperature and shivering fits. The fevers can occur many times, frequently resulting in the death of the victim. Malaria is a particularly big killer of children, who have not had time to build up any immunity to the disease. Even if malaria does not kill the victim it can cause kidney failure and leaves the patient weak, anaemic and prone to other diseases. The infected person's life expectancy is reduced considerably.

**Did you know...?** Every 60 seconds, a child dies from malaria.

# MALARIA – ITS CAUSE AND TRANSMISSION

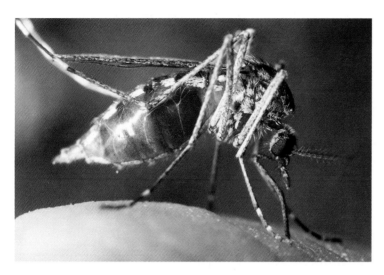

**Figure 29.1**
A mosquito sucking up blood

In areas where malaria occurs, many of the people will have the disease. As a result, the amount of wealth (from farms and factories) that the area produces is seriously reduced while, at the same time, a lot of time and money has to be spent on caring for victims of the disease. In the Philippines, for example, malaria was rife in the 1940s and as a result 35% of people were unable to work. In regions where malaria is particularly bad, for example northern Sri Lanka, people have been forced to move away, leaving behind fertile farmland.

The parasite that causes malaria enters a person's bloodstream when he or she is bitten by a mosquito. Not all mosquitoes carry the disease. **Only the female *Anopheles* mosquito spreads malaria.** It bites an infected person and sucks blood containing the parasite into its stomach, where the parasites multiply. The mosquito then bites someone else and the parasite enters the new victim via the saliva of the mosquito.

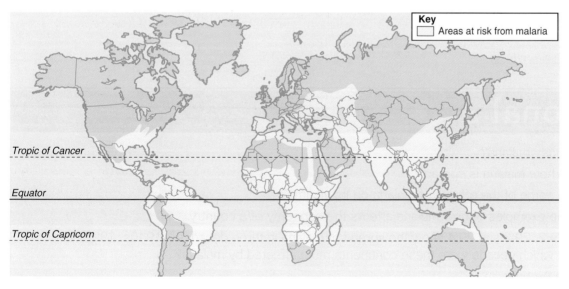

**Figure 29.2**
Distribution of malaria (1940s)

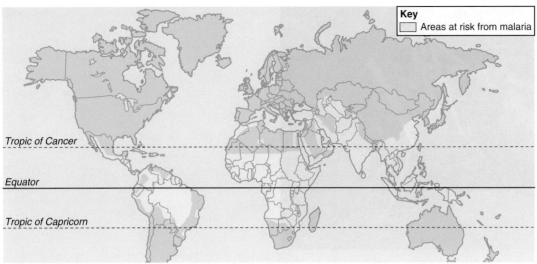

**Figure 29.3**
Distribution of malaria (1990s)

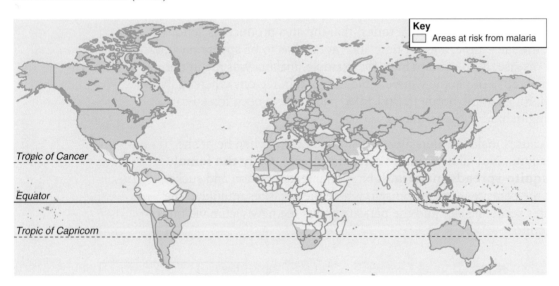

**Figure 29.4**
Distribution of malaria (2010s)

# National 4

1. What causes malaria?
2. Describe how malaria is spread.
3. What are some of the effects of malaria on its victims?
4. Give three examples of how malaria affects the economy of a country.
5. Figures 29.2, 29.3 and 29.4 show the spread of malaria in three decades – 1940s, 1990s and 2010s. In which decade were these continents most affected by malaria?
    (a) North America
    (b) Africa
    (c) Asia
6. Overall, how has the global distribution of malaria changed since the 1940s?

# MALARIA – ITS CAUSE AND TRANSMISSION

## National 5

1. Describe in detail the causes of malaria and how it is spread.
2. Give examples of some of the effects of malaria on its victims.
3. Explain how malaria slows down the economic development of a region.
4. Look at Figures 29.2 and 29.3. In which continents had malaria increased between the 1940s and 1990s, and in which continents had it decreased?
5. Look at Figures 29.3 and 29.4. In which continents had malaria increased between the 1990s and 2010s, and in which continents had it decreased?
6. Look at Figure 29.4 and describe the global distribution of malaria in the 2010s.

## Activity

| Country | 1940s | 1990s | 2010s |
|---|---|---|---|
| Australia | SOME | NONE | NONE |
| Cuba | | | |
| Italy | | | |
| Kazakhstan | | | |
| Madagascar | | | |
| Paraguay | | | |
| Somali Republic | | | |
| South Korea | | | |
| Spain | | | |
| Turkey | | | |
| Venezuela | | | |
| Yemen | | | |

**Table 29.1**
Countries with malaria

(a) Copy Table 29.1 into your notebook.
(b) Using an atlas and Figures 29.2, 29.3 and 29.4, complete the table. For each time period, you must write down whether ALL, SOME or NONE of the country had malaria. The first one has been done for you.

**Now complete the 'I can do' boxes for this chapter.**

# Chapter 30

## Malaria – factors in its distribution

This chapter looks at the factors affecting the spread of malaria.

**By the end of this chapter, you should be able to:**

- ✓ describe the physical environment where malaria is likely to be found
- ✓ explain the human factors that contribute to the spread of malaria
- ✓ give examples of methods used to control malaria.

## Factors in the distribution of malaria

Malaria has occurred in most areas of the world at some time. As Figure 29.2 shows, in the 1940s even people in the USA and Europe were affected. The disease then retreated (Figure 29.3) and the most recent map of its distribution (Figure 29.4) shows little change from the 1990s. Today, malaria exists in 100 countries and over 60% of the world's population live in malaria-affected areas. The reasons behind its changing distribution are a combination of physical and human factors.

### Physical factors

**Malaria occurs in areas where the *Anopheles* mosquito lives.** These mosquitoes live in both warm and hot areas, where **temperatures are above 16 °C**. They need **still water surfaces** as breeding areas, but these areas do not need to be large. As a result, all **warm, rainy areas** with still or slow-moving water are suitable environments for the *Anopheles* mosquito.

# MALARIA – FACTORS IN ITS DISTRIBUTION

## Human factors

People's activities also affect the distribution of malaria. **Where people have built dams and made irrigation channels**, they have created suitable breeding areas for the mosquito and so the incidence of malaria increases. **People migrate much more now** and this makes it easier for the disease to spread. In some areas of the world, people have successfully used insecticides to kill off the mosquito. This explains why the areas affected by malaria decreased between the 1940s and the 1990s.

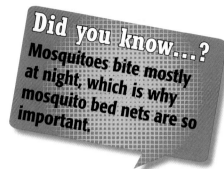

Did you know...?
Mosquitoes bite mostly at night, which is why mosquito bed nets are so important.

## Controlling malaria

**Malaria can be both prevented and cured by anti-malarial drugs** such as Chloroquine, Larium and Malarone, which kill the parasite that causes the disease. However, the malarial parasites can develop resistance to anti-malarial drugs, so the drugs have to be constantly re-developed in order to be effective. Instead of curing people, it is much better and cheaper to prevent people from catching malaria by reducing their exposure to mosquitoes. **To prevent malaria, the mosquitoes must be destroyed.** Many methods of prevention have been tried (see Table 30.1) but they all have drawbacks, which is why so little progress has been made in recent years.

Did you know...?
Malaria is preventable, treatable and curable, yet it is the largest killer of children in the world.

| Method of prevention | Drawbacks |
| --- | --- |
| Use insecticides (e.g. DDT and Malathion) | Some chemicals pollute the environment, killing other life forms<br><br>Mosquitoes are becoming resistant to some insecticides<br><br>Some insecticides are expensive |
| Use insecticide-treated bed nets to prevent being bitten at night | Developing countries cannot afford to buy bed nets |
| Drain breeding grounds | Impossible to drain all breeding grounds, as only small areas of water are needed for larvae to exist, e.g. potholes in roads |
| Kill the mosquito larvae (e.g. using egg whites, coconuts, mustard seeds and fish in the water) | This is wasteful of food; it is also wasteful of water |
| Educate people, so they know how the disease is spread and can reduce their contact with mosquitoes | People still have to live near water, which is where mosquitoes live |

Table 30.1
Methods of preventing malaria

GLOBAL ISSUES

## National 4

1. Describe the physical environment in which *Anopheles* mosquitoes live.
2. In what ways have people helped to spread malaria?
3. Describe the ways in which malaria can be prevented.
4. Choose one of the methods of prevention shown in Table 30.1 and explain whether you think it is effective or not.
5. Describe how malaria can be cured.

## National 5

1. Describe, in detail, the physical environment in which malaria is found.
2. Explain how people have helped to spread malaria.
3. Describe, in detail, the ways in which malaria can be prevented.
4. Look at Table 30.1. Which are the two best methods of preventing malaria? Explain your decision.
5. Explain, in detail, why little progress has been made in recent years in reducing the spread of malaria.

## Activities

### Activity A

Study the image below carefully.

(a) Do you think the people in the photo suffer from malaria?
(b) What other information would help you to answer this question?

# MALARIA – FACTORS IN ITS DISTRIBUTION

## Activities continued...

### Activity B

Imagine you are a voluntary worker who has gone to a village to warn the people living there of the dangers of malaria and to explain how they can reduce the risk of contracting the disease. Make an information poster informing the villagers of what they can do to reduce their chances of contracting the disease.

### Activity C

'There is much more research into male baldness than there is into diseases such as malaria.'

(Bill Gates)

Look at the statement above by Bill Gates, co-founder of Microsoft. Write down what you think about this statement.

Now complete the 'I can do' boxes for this chapter.

# Chapter 31

## Heart disease – its causes

This chapter looks at a common disease in rich countries – heart disease.

**By the end of this chapter, you should be able to:**

✓ describe the main causes of heart disease
✓ draw and interpret graphical data.

In developed countries, heart disease is the second biggest cause of death after cancer. It kills nearly half of all men and women. One in four men will have a heart attack before retirement age and most teenagers show signs of narrowing of the blood vessels, which is the start of heart disease. Unlike major diseases in developing countries, heart disease is non-infectious – one person cannot infect another. Instead, the causes of heart disease are more complicated.

## Causes of heart disease

Heart (cardiovascular) disease can cause stroke, angina and heart attacks. Some types of heart disease affect the arteries (which carry blood from the heart to the rest of the body); others affect the heart itself. Many factors contribute to heart disease.

# HEART DISEASE – ITS CAUSES

## Poor diet
**Too many fatty foods increase cholesterol**, which is a type of fat found in the blood. This narrows the arteries, increasing the chance of heart disease. Fatty and sugary foods also lead to a person becoming obese or overweight, which puts an extra strain on their heart.

## Lack of exercise
**Lack of exercise raises blood pressure and cholesterol** levels and can also cause a person to become overweight.

## Smoking
**Nicotine increases the heart rate and blood pressure**, so more oxygen is needed for the heart to work properly. However, smokers receive less oxygen while smoking, putting their heart under extra strain. A packet of cigarettes a day doubles a person's chances of having a heart attack and makes them five times as likely to have a stroke.

## Stress
**Stress increases a person's blood pressure** and this puts extra pressure on their heart. People under stress often indulge in 'comfort eating', for example chocolate bars or greasy chips, which can also cause heart disease.

## Inheritance
**People can inherit a risk of high blood pressure and high cholesterol** levels from their parents.

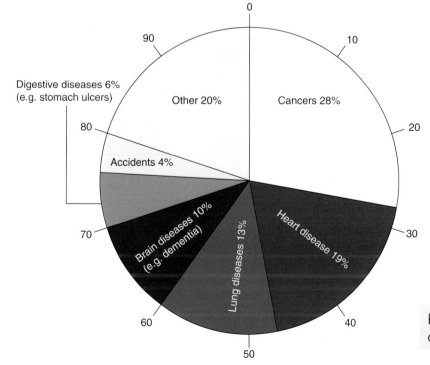

Pie charts should be used to compare parts of a whole.

**Figure 31.1**
Causes of death in Scotland (2016)

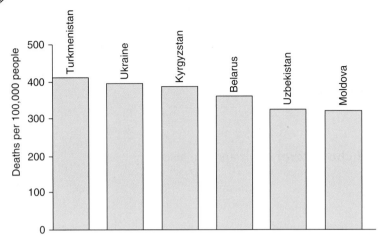

**Figure 31.2**
Countries with the highest death rate from heart disease (2017)

Bar charts should be used to compare separate amounts.

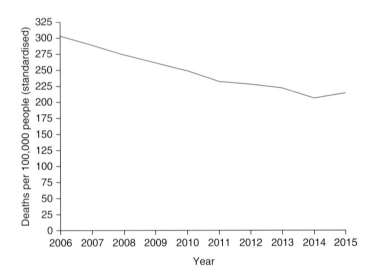

**Figure 31.3**
Changes in death rate from heart disease in Scotland (2006–2015)

Line graphs should be used to show changes over distance or time.

| Type of cost | Amount (£ billion) |
| --- | --- |
| Prevention | 0.5 |
| Primary care (e.g. work done by GPs) | 1.7 |
| Outpatient care at hospitals | 1.6 |
| Inpatient care (hospital admissions) | 9.8 |
| Medicines | 3.2 |
| Rehabilitation (recovery needs) | 0.8 |
| Social services (care at home) | 2.4 |
| **Total** | 20.0 |

**Table 31.1**
Cost of health care for heart disease in the UK (2015)

# HEART DISEASE – ITS CAUSES

| Type of cost | Amount (£ billion) |
|---|---|
| Health care | 20.0 |
| Productivity loss (loss of production because people cannot work or because they die young) | 13.9 |
| Other care costs (e.g. family members looking after those with heart disease) | 10.1 |
| **Total** | 44.0 |

Table 31.2
Total cost of heart disease in the UK (2015)

| Region | Death rates from heart disease per 100,000 people (standardised) | | | |
|---|---|---|---|---|
| | 2006 | 2009 | 2012 | 2015 |
| Greater Glasgow | 310 | 268 | 237 | 222 |
| Lothian | 289 | 249 | 216 | 199 |
| Grampian | 290 | 251 | 234 | 211 |
| Tayside | 260 | 248 | 227 | 212 |
| Borders | 302 | 278 | 227 | 243 |
| Highlands | 288 | 232 | 209 | 215 |

Table 31.3
Death rates from heart disease in regions of Scotland (2006–2015)

## National 4

| Cause of heart disease | Explanation of the cause |
|---|---|
| Poor diet | |
| Lack of exercise | |
| Smoking | |
| Stress | |
| Inheritance | |

1. Copy the table above into your notebook and complete it using the information in this chapter.
2. Look at Figure 31.1. What percentage of people in Scotland die from:
   (a) heart disease
   (b) cancer?
3. Look at Figure 31.2. What is the death rate from heart disease in:
   (a) Moldova
   (b) Belarus?
4. Look at Figure 31.3 and describe the changes in deaths from heart disease in Scotland since the year 2006.

## National 4 continued...

5. Look at Table 31.3.
   (a) Between 2006 and 2015, in which region did death rates from heart disease:
      (i) fall the most
      (ii) fall the least
   (b) Draw the most appropriate graphs (line, bar or pie) to show these changes.
6. What would be the most appropriate graph (line, bar or pie) to show the information in Table 31.1? Give a reason for your answer.
7. Table 31.2 shows the costs of heart disease to the UK economy. Choose one of the items in this table and give details of the costs involved.

## National 5

1. Describe the effects of smoking on your heart.
2. Explain how eating fatty foods increases your chances of developing heart disease.
3. Explain how obesity can lead to heart disease.
4. Heart disease can be caused by high blood pressure. What can cause people to have high blood pressure?
5. Describe what is shown by Figure 31.1.
6. Look at Figure 31.2. Compare the death rates from heart disease in Moldova, Turkmenistan and Belarus.
7. Look at Figure 31.3. Describe in detail the changes in deaths from heart disease in Scotland since the year 2006.
8. Explain what is shown by Table 31.2. You may wish to use the information in Table 31.1 in your answer.
9. Look at Table 31.3.
   (a) Which three regions reduced their death rate from heart disease the most between 2006 and 2015?
   (b) Draw an appropriate graph(s) to show these changes.
10. What would be the most appropriate types of graph (line, bar or pie) to show the information in Tables 31.1 and 31.2? Give reasons for your answer.

# HEART DISEASE – ITS CAUSES

## Activity

> Scotland has been branded the heart attack capital of Western Europe. This is caused largely by

> Cases of heart disease finally starting to fall in Scotland. Doctors put it down to

> Research claims heart disease was much lower in Scotland 100 years ago. Scientists say this was due to

> New report says it should be easy to reduce heart disease in Scotland. We just need to

The above newspaper headlines have been torn off. Write them out and complete what you think they were going to say.

**Now complete the 'I can do' boxes for this chapter.**

# Chapter 32

## Heart disease – methods of control

*This chapter looks at the distribution of heart disease and ways of controlling it.*

**By the end of this chapter, you should be able to:**

- describe the factors in the distribution of heart disease
- give examples of methods of controlling heart disease
- explain the role played by the National Health Service in preventing and treating heart disease.

## Factors in the distribution of heart disease

The countries most affected by heart disease are all developed countries, but there are big differences between them. Scotland has one of the worst heart disease rates in the world, but even here some areas (e.g. Glasgow) are much worse than other areas. The main reasons why heart disease varies so much from one area to another and from one developed country to another are described below.

### Lifestyle

**In developed countries** and in cities throughout the world, **the pace of life is faster**. Many more people work in offices and take little exercise.

### Diet

**Some countries have healthier diets than others.** For instance, the traditional Asian diet contains very little meat and dairy produce and Japanese people have a much lower heart disease rate than British people. A Mediterranean diet also contains fewer saturated fats and people there have lower rates of heart disease too.

## Affluence

**People in richer countries can afford to eat too much** and can afford to buy cigarettes and alcohol, so are more likely to develop heart disease. Within richer countries, **the cheapest foods are often fatty foods** and so poorer people are more likely to develop heart disease.

## Medical care

Developed countries, for example Australia, have run very good **campaigns to educate people on how to reduce heart disease**. As a result, the incidence of heart disease has reduced dramatically in recent years.

Within any country, **the number of people with heart disease depends on the treatment available** locally, for example how well the local health authorities try to diagnose and prevent heart disease and the equipment and drugs available there to treat the disease.

# Controlling heart disease

The death rate from heart disease in the UK has been dropping in recent years; it dropped by 44% between 2000 and 2010. Some countries, such as Australia, Canada and Sweden, have done even better than this. The death rate has dropped because of better prevention and better treatment.

In the UK, **the National Health Service (NHS) helps to control heart disease** by providing free medical check-ups, giving advanced treatment and educating people.

## Medical check-ups

More people now have regular cholesterol and blood pressure check-ups, which enables them to find out if they are at risk of heart disease and to take action before it is too late.

## Advanced treatment

Better medical equipment is being invented and used in the treatment of heart disease, including pacemakers, artificial heart valves and defibrillators. The success rate of heart bypass surgery is steadily improving and more drugs are being developed, for example aspirin to reduce blood clotting, beta-blockers to reduce heart rate and alpha-blockers to reduce blood pressure.

## Education

People are being educated through campaigns in the media and on posters. Money spent on preventing people developing heart disease is money well spent, as the cost of treating them for heart disease is much greater. The four main pieces of advice given by the NHS are shown below.

| Eat less | Eat more |
|---|---|
| full milk and cream | skimmed milk |
| butter | polyunsaturated margarine |
| fried food | grilled food |
| milkshakes | low-calorie soft drinks |
| sausages, pork pies | chicken, turkey |
| cakes, biscuits, sweets | oats, pasta, cereals |
| chips, crisps | fruit and vegetables |
| white bread | brown bread |

**Figure 32.1**
Preventing heart disease

Did you know...? Fat is not the biggest cause of diet-related heart disease, sugar is!

1. **Eat a better diet.** Advice is given on foods that are healthy and which foods should be reduced or avoided (see Figure 32.1). Food labels now contain much more information. Despite these measures, the number of overweight and obese children in Scotland is increasing, from 27% in 2000 to 29% in 2016.
2. **Take more exercise.** People are encouraged to take more exercise and sports facilities have increased, for example there are now more jogging tracks, cycle lanes, gyms and sports centres. As a result, evidence suggests that the average person now takes more exercise.
3. **Stop smoking.** There have been extensive campaigns to persuade people to stop smoking, and there is more help available than ever before, for example nicotine patches, helplines and hypnotism. The number of smokers is now fewer than 20 years ago.
4. **Reduce stress levels.** People are now more aware that stress is harmful and understand ways to reduce stress, for example relaxing by taking exercise or listening to music. But there is no evidence that stress levels are decreasing. In fact, it is more likely that they are increasing.

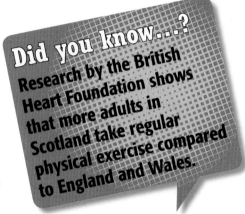

Did you know...? Research by the British Heart Foundation shows that more adults in Scotland take regular physical exercise compared to England and Wales.

Did you know...? Surveys show that more girls than boys are regular smokers.

# National 4

1. Name four factors affecting the distribution of heart disease.
2. How does lifestyle affect the distribution of heart disease?
3. How does affluence affect the distribution of heart disease?
4. (a) What advice is given to people to prevent heart disease?
   (b) Are people in the UK taking this advice?
5. How does the NHS try to control heart disease?

# HEART DISEASE – METHODS OF CONTROL

## National 5

1. Describe, in detail, three factors that affect the distribution of heart disease.
2. Explain clearly the ways that heart disease could be prevented.
3. The NHS tries to control heart disease by educating people.
   (a) Describe the advice people are given.
   (b) Describe how successful this advice is.
   (c) Describe two other ways in which the NHS controls heart disease.

## Activities

### Activity A

| Health indicator | Year 2000 | Year 2010 |
|---|---|---|
| Smokers (adults) | 28% | 21% |
| Taking enough physical exercise (male) | 36% | 42% |
| Taking enough physical exercise (female) | 23% | 31% |
| Heavy drinkers (female) | 10% | 13% |
| Heavy drinkers (male) | 21% | 19% |
| Low-calorie soft drinks (ml per person per week) | 516 | 579 |
| Skimmed milk (litres per week) | 1.16 | 1.16 |
| Sugar (grams per person per week) | 130 | 90 |
| Salt (grams per person per week) | 9 | 11 |
| Butter (grams per person per week) | 37 | 40 |
| Low-fat spread (grams per person per week) | 22 | 11 |
| Chocolate bars (grams per person per week) | 113 | 89 |
| Fruit and vegetables (grams per person per week) | 2240 | 2381 |

Table 32.1
Comparison of health indicators in 2000 and 2010

You have been asked to give a report to the Minister of Health on how healthily we live in the UK.

(a) Look at Table 32.1. Using bullet points, list the four main points that you would make to the Minister.
(b) Describe your main recommendations. What are the two main problems and how do you think they would best be solved?

## Activities continued...

### Activity B

Choose one of the methods of controlling heart disease listed below.

(a) Write down what you think would be the best way of persuading teenagers to adopt the method and why.
(b) Draw a poster with an eye-catching slogan to get this message across.

- Eat a better diet
- Stop smoking
- Take more exercise
- Reduce stress levels

**Now complete the 'I can do' boxes for this chapter.**

# Chapter 33

This chapter looks at HIV/AIDS.

# HIV/AIDS – its distribution

By the end of this chapter, you should be able to:

- ✓ explain how HIV/AIDS affects people
- ✓ describe the global distribution of HIV/AIDS
- ✓ draw and interpret choropleth maps.

**Did you know...?**
HIV/AIDS is the world's leading infectious killer.

**Did you know...?**
In Africa, because of HIV/AIDS, life expectancy is now lower than it was 30 years ago.

AIDS (acquired immunodeficiency syndrome) is caused by a virus called HIV (the human immunodeficiency virus). If you are infected with HIV, your body tries to fight the infection by making antibodies. A person who is HIV-positive has these antibodies inside them, which means they have the HIV virus.

The HIV virus gradually wears down a person's immune system, making it more and more difficult for them to fight disease. This means that viruses and bacteria (which do not cause ill-effects in healthy people) make people who are HIV-positive very sick. They develop a group of health problems (a syndrome) called AIDS and eventually, because they have little resistance to disease, HIV-positive people will die.

## Distribution of HIV/AIDS

In 2010 there were 34 million people worldwide living with HIV/AIDS. Every country in the world has cases, but the overall distribution is very uneven (see Table 33.1).

Two-thirds of all cases of HIV/AIDS are in Africa but the distribution there is also very uneven. Figure 33.2 is a choropleth map showing the cases of HIV/AIDS in Africa.

The countries are shaded differently according to the total number of cases found there – the darker the shading, the greater the number of cases. The map clearly shows the uneven distribution.

| Region (see Figure 33.1) | Number of people with HIV/AIDS (million) | Percentage of all adults with HIV/AIDS (%) |
| --- | --- | --- |
| Sub-Saharan Africa | 22.9 | 5.0 |
| North Africa and Middle East | 0.47 | 0.2 |
| South and Southeast Asia | 4.0 | 0.3 |
| East Asia | 0.79 | 0.1 |
| Oceania | 0.05 | 0.3 |
| Latin America | 1.5 | 0.4 |
| Caribbean | 0.20 | 0.9 |
| Eastern Europe and Central Asia | 1.5 | 0.9 |
| North America | 1.3 | 0.6 |
| Western and Central Europe | 0.84 | 0.2 |
| WORLD TOTAL | 34 | 0.8 |

**Table 33.1**
People living with HIV/AIDS (2010)

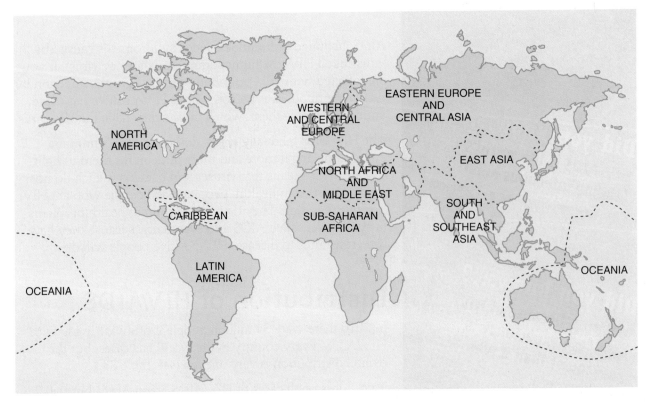

**Figure 33.1**
Regions of the world referred to in Table 33.1

# HIV/AIDS – ITS DISTRIBUTION

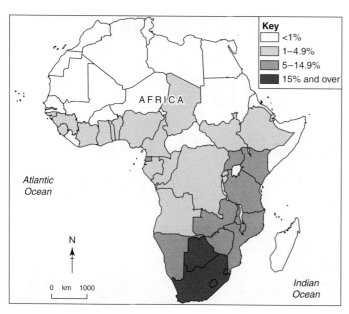

**Figure 33.2**
Percentage of people affected by HIV/AIDS in African countries, 2012

## National 4

1. Describe how HIV affects its victims.
2. Look at Table 33.1.
   (a) Which two regions have the most cases of HIV/AIDS?
   (b) How many people have HIV/AIDS in these two regions?
   (c) How many people in the world have HIV/AIDS?
3. Look at Figure 33.2 and a map of Africa.
   (a) Name three countries where HIV/AIDS is less than 1%.
   (b) Name three countries where HIV/AIDS is greater than 15%.
4. (a) Mark the regions shown in Figure 33.1 on an outline map of the world.
   (b) Draw a choropleth map to show the percentage of adults with HIV/AIDS in these different regions, using the categories given by your teacher.

## National 5

1. Describe in detail how HIV/AIDS affects the victim.
2. Using Table 33.1, describe the global distribution of HIV/AIDS.
3. (a) Mark the regions shown in Figure 33.1 on an outline map of the world.
   (b) Draw a choropleth map to show the percentage of adults with HIV/AIDS in these different regions. You will first need to decide how many categories you wish to have and what the categories will be.

Now complete the 'I can do' boxes for this chapter.

# Chapter 34

*This chapter looks at factors affecting the distribution of HIV/AIDS.*

# HIV/AIDS – causes, effects and treatment

**By the end of this chapter, you should be able to:**

- ✓ describe the factors affecting the distribution of HIV/AIDS
- ✓ explain the consequences of HIV/AIDS for a country
- ✓ give examples of how HIV/AIDS can be treated.

## Factors in the distribution of AIDS

When the blood or body fluids of a person with HIV/AIDS are passed on to someone else, that person also becomes infected. **The main ways in which people contract HIV/AIDS are:**

- **sharing a needle with an infected person**
- **having unprotected sex with an infected person**
- **babies drinking the breast milk of an infected mother.**

Once a person is HIV-positive, other factors which weaken their immune system make him or her more likely to develop AIDS. **The main weakening factors are:**

- drug abuse
- poverty
- malnutrition
- depression
- other infections.

In poorer countries, especially in Africa, the greater number of cases of AIDS is due to the greater number of people living in poverty and suffering from malnutrition and other

# HIV/AIDS – CAUSES, EFFECTS AND TREATMENT

infections. In addition, there is less health education so many people are unaware of the causes of AIDS and how the risk of infection can be reduced. **War and the breakdown of law accelerate the spread of AIDS** because of the poverty and malnutrition which follows, and because of the higher incidence of rape. Sadly, many areas of sub-Saharan Africa have suffered conflict in recent years, for example Sudan and Democratic Republic of the Congo.

**In developed countries, drug abuse is a much bigger factor in the spread of AIDS** than it is in developing countries.

Did you know…? More than 6800 new cases of HIV occur each day worldwide.

## Consequences of HIV/AIDS

In 2010, 69% of the people with HIV/AIDS lived in Africa and in some African countries one in every three adults was infected. The consequences for these countries are extremely serious.

- **The prevention, detection and treatment of AIDS is expensive.** To treat all the AIDS-infected people in Africa would cost three times the amount of money available for all health care. In some African countries, over one-third of all hospital beds are now occupied by HIV/AIDS patients. **To effectively treat AIDS would mean less money and less health care for people with other diseases.**
- As adults become ill, **responsibility falls on the older children to try and earn money, provide food and care for their family**. This is almost impossible for them and is often at the expense of their own education. So **the next generation of African adults will be less educated, less wealthy and less healthy** than the previous generation.
- Table 34.1 shows the effect of AIDS on life expectancy in a sample of African countries. **People are dying younger** and so have fewer years when they are economically active. In addition, as they fall ill **they are less able to work.** With so many adults affected, **this is seriously reducing production on farms, in factories and in offices** in every African country. South Africa has calculated that its total income will reduce by nearly 20% because of AIDS. Because there are fewer people working, **there are fewer taxpayers, so the country is producing less wealth** and also has less tax money to pay for services and to carry out development plans.

## Treatment

There is **no cure for AIDS** but there are ARV (antiretroviral) **drugs which slow down the effects of the HIV virus**. There are also **drugs which stop the disease passing from pregnant mothers to their babies. Health education programmes** can be used to try to prevent the disease from spreading. It is also important for people to be tested, as many do not know they are carrying the disease. In addition, efforts can be made to **reduce the effects of factors which hasten the development of AIDS**. Attempts to reduce poverty and improve diet should therefore reduce malnutrition and slow down the progress of the disease.

| Country | Life expectancy before AIDS | Life expectancy in 2010 |
|---|---|---|
| Botswana | 74.4 | 53.3 |
| Lesotho | 67.2 | 46.0 |
| Malawi | 69.4 | 51.5 |
| Namibia | 68.8 | 61.9 |
| Rwanda | 54.7 | 53.9 |
| South Africa | 68.5 | 51.2 |
| Swaziland | 74.6 | 47.3 |
| Zambia | 68.6 | 46.9 |
| Zimbabwe | 71.4 | 46.5 |

Table 34.1
Changes in life expectancy in nine African countries

### Case study: South Africa

South Africa has more people with HIV/AIDS than almost any other country in the world (one in every six) but, before 1982, there were none. By 1990, 2% of all adults in the country were infected and, by 2000, this had increased to 20%. In 2010 the infection rate had dipped to 18% but 800 people were still dying each day from the disease.

Although drugs are available to treat AIDS, in South Africa there are **not enough trained staff to administer the treatments** and there are **many very isolated areas** which are difficult to reach. Many people do not know they have the disease. **Testing facilities are poor** and many **people avoid testing** because of the stigma associated with the disease.

In 2000, the government began a big recruitment and training programme for medical staff, and ARV drugs were made available much more cheaply (£100 per patient per year). In addition, 160 million free condoms were distributed.

Since 1998, there have been **HIV education campaigns**, informing people about AIDS and the need to practise safe sex. This has been made difficult because **one in seven South Africans cannot read** and there are **eleven official languages** in the country. To overcome this, the campaign uses radio, TV soap operas and drama. None of South Africa's strategies for dealing with AIDS began until the disease was already rife in the country and this has made it much more difficult to control.

> **Did you know...?**
> A study in 2004 found that South Africans spend more time attending funerals than shopping.

## National 4

1. What are the three main ways in which people contract HIV/AIDS?
2. Why is AIDS more common in developing countries than in developed countries?
3. Why does AIDS spread more quickly in countries at war?
4. Choose one of the consequences of HIV/AIDS and explain it.
5. Describe the ways in which AIDS can be treated.
6. List four problems in trying to treat AIDS in South Africa.

## National 5

1. Explain fully why AIDS is more common in developing countries than in developed countries.
2. AIDS spreads more quickly in countries at war. Explain why.
3. In a country in which many people suffer from HIV/AIDS, explain why:
   (a) the children are less educated
   (b) there is less tax money
   (c) there is less wealth produced.
4. Describe in detail the ways in which AIDS can be treated.
5. Describe three problems in treating AIDS in South Africa.

# Activities

## Activity A

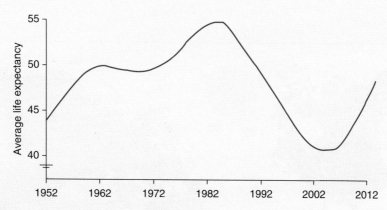

The graph above shows changes in life expectancy in a typical country in southern Africa over the last 50 years.

(a) Copy the graph.
(b) Decide when the following events took place in the country shown and write them in the correct place on the graph:

- ARV drugs available
- Free primary education introduced
- First case of HIV/AIDS
- Short civil war
- Free condoms available
- New water supply schemes
- Trade deal with the EU
- Drought
- Poverty increasing rapidly
- MP says garlic cures AIDS

## Activity B

Below is a completed crossword but with the clues missing. You must come up with a clue for each of the words in the crossword to describe what each word means in relation to HIV/AIDS.

|   |   |   |   |   | ¹B |   |   | ²P |
|---|---|---|---|---|---|---|---|---|
|   | ³M | ⁴A | L | N | U | T | R | I | T | I | O | N |
|   |   | I |   |   | E |   |   | V |
| ⁵N | E | E | D | L | E |   |   | E |
|   |   | S |   |   | A |   |   | R |
|   |   |   |   |   | S |   |   | T |
|   |   |   | ⁶T | R | E | A | T | M | E | N | T |
|   |   |   |   |   |   |   |   | Y |
|   |   | ⁷I | N | F | E | C | T | I | O | N | S |
|   |   |   |   |   | L |   |
|   |   |   |   |   | K |   |

**Now complete the 'I can do' boxes for this chapter.**

# N5 Examination questions

Read the information and the two SQA-style questions below, and answer Tasks A–H.

## Question 1

**Figure 34.1**
Diagram Q1 – Distribution of men with lung cancer (2012)

(a) Study Diagram Q1. Describe, in detail, the distribution of men with lung cancer in 2012. (4)

(b) Choose one of these three diseases: **heart disease; cancer; asthma**. For the disease you have chosen, explain the methods used to control it. (6)

### Advice for Question 1(a)

You should have spotted four important points.

1 It is a *describe* question.

2 It has a diagram – Diagram Q1. The question is testing your skill in describing the diagram, a world map. We have already discussed this type of question in previous chapters.

### Advice for Question 1(b)

You should have spotted four important points.

1 It is an *explain* question, which is the most commonly asked question type for part (b). So you need to make a number of points which make it clear how your chosen disease can be controlled.

2 It has no diagram. You must rely on your own knowledge and understanding.

# N5 EXAMINATION QUESTIONS

## Advice for Question 1(a)

3 It tells you to *describe the distribution*. This is a common question asked. You need to describe the pattern shown – where are the most and fewest cases of lung cancer? You should name regions (countries, if possible) and spot any patterns – for example, are there more in cold or hot areas, in developed or developing countries, in industrial or farming regions?

4 It is worth **4** marks, so you need to make 4 factual points or at least 2 developed points. To make a developed point, you could use the key and mention figures and/or you could refer to specific countries from your knowledge. For example, *There are few cases of lung cancer in central Africa. Countries such as Sudan and DR Congo have fewer than 4.3 cases of lung cancer per 100,000 men.* (2)

## Advice for Question 1(b)

3 You must *choose* **one** disease from the three given, and you should choose the one that you know best.

4 It is worth **6** marks, with developed points earning 2 marks. For example, *Asthma can be controlled in the short-term with inhalers.* (1) *It can be controlled in the longer-term with medicines which reduce the inflammation of the airways and help the patient to breathe more easily.* (2)

**TASK A**: Read the advice and then answer Question 1(a).

**TASK B**: Read the advice and then answer Question 1(b).

## Question 2

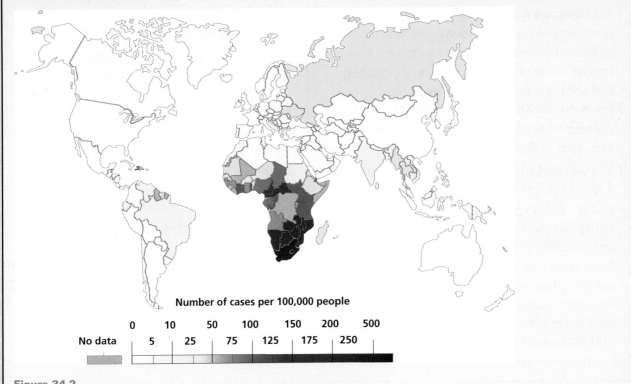

**Figure 34.2**
Diagram Q2 – Deaths from HIV/AIDS per 100,000 people (2016)

**(a)** Study Diagram Q2. Describe, in detail, the distribution of deaths from HIV/AIDS. (4)

**(b)** Explain the impact of HIV/AIDS on the people and countries affected. (6)

### How good is this answer to Question 2(a)?

North America has few deaths from HIV/AIDS, fewer than 10 people. South America is slightly worse with some countries up to 25. Europe is the same as North America, but Asia is slightly worse than South America because some countries are as high as 50. Australia is the same as Europe but Africa is the highest of all. Some countries have up to 500 deaths.

**TASK C**: Read the answer above and give it a mark. Explain the number of marks you have given.

### How good is this answer to Question 2(b)?

Children with HIV/AIDS are much less likely to go to school and so they will have poor job prospects. They are likely to be too unwell to work at all. Their health will be poor, they will need medicine and they will spend much time travelling to and from clinics. This is very costly, so their family has less money for clothing, food and school fees. The people infected might well be shunned by the local community, which could cause them to be depressed.

**TASK D**: Read the answer above and give it a mark. Explain the number of marks you have given.

## Question 1 on page 184

### This is a good answer to Question 1(a)

Developed countries, such as the USA and Canada, have the highest lung cancer rates, over 41.5 cases per 100,000 men. (2) Middle-income countries, such as Brazil and Argentina and South Africa, have lower rates but still more than 12 per 100,000. (1) The poorest countries in the world have the fewest cases. (1) Countries such as Ethiopia, Sudan and Somalia all have less than 4.3 cases per 100,000. (1)

### Why this is a good answer

The candidate has spotted the connection between economic development and lung cancer cases. They have developed each point (each sentence) by giving examples of countries and using the scale to give the number of cases. Three developed points is more than enough to score full marks.

**TASK E**: The answer is very good but there are also several exceptions to the trends described which could have earned the candidate more marks. Rewrite the answer, including exceptions. For example, *Developed countries have the highest lung cancer rates except for ...*

### This is a good answer to Question 1(b)

I have chosen heart disease and I am using South Korea as an example. It has controlled heart disease very well by setting up special heart hospitals all over the country, which have specialist surgeons and the most up-to-date equipment. (2) Nearly everyone goes for a check-up because there are lots of medical centres and people have health insurance, (2) so they know they can be treated for free. (1) There are lots of public health messages from the government which people can get on their TVs and phones, and nearly everyone has access to these. (1)

### Why this is a good answer

The candidate has given their choice at the start. The question does not tell you to write about specific countries but it is OK to do so. The candidate has made it clear how the disease is being controlled, and they have made several points including 2 (possibly 3) developed points.

**TASK F**: Improve this answer by giving examples of the public health messages the South Korea government will give (which will be similar to those given in other countries).

## Question 2 on page 186

### Advice for Question 2(a)

You should have spotted three important points.

1. It is a *describe* question and it has a diagram – Diagram Q2. You are expected to interpret the map and state a few facts on the distribution of HIV/AIDS deaths.
2. It asks you to *describe the distribution*, so you must try to give facts about the location and the overall pattern.
3. It is worth **4** marks, so you must make 4 valid points or 2 developed points. To make developed points, refer to specific countries and mention figures by interpreting the scale.

### Advice for Question 2(b)

You should have spotted four important points.

1. It is an *explain* question, so you must make the effects of HIV/AIDS clear.
2. It has no diagram. You are expected to use your own knowledge and understanding of this topic.
3. It is a two-part question; you need to explain the impact on both individuals and the whole country.
4. It is worth **6** marks, so it helps to try to make detailed points worth 2 marks.

### How good was the answer on p.186?

(Mark: 2 out of 4)

It is easy to see the candidate's approach – go around the world map, commenting on each continent. Strictly, they have described the location more than the distribution of HIV/AIDS deaths. Location refers to the individual places; distribution refers to the overall pattern. To describe the distribution, the candidate could state that there are far more deaths in developing countries than developed. There are also more in the southern hemisphere than in the north. It is a good idea to give actual figures but you must take care in reading the scale. North America does not have 10 deaths but 10 deaths for every 100,000 people.

**TASK G**: Read the comments above and then write an improved answer to Question 2(a).

### How good was the answer on p.186?

(Mark: 4 out of 6)

This candidate has given several effects of HIV/AIDS on people, making 2 developed points and 2 other points as well. They have not just stated the impact but explained why (they will have less money, poor job prospects, etc.) – just as the question asks. But the question also asks you to explain the impact on the country (for example, fewer taxpayers, fewer workers). Because this candidate did not do this, the maximum mark they achieve is 4.

**TASK H**: Read the comments above and then write an improved answer to Question 2(b).

# 'I can do' checklist

|  | Red | Yellow | Green | Comment |
|---|---|---|---|---|
| **Chapter 1 Climate change** | | | | |
| Describe what climate change is | | | | |
| Describe the natural greenhouse effect | | | | |
| Give examples of evidence that our climate has changed | | | | |
| **Chapter 2 Climate change – physical factors** | | | | |
| Describe how physical factors affect the world's climate | | | | |
| Give examples of some physical factors that cause climate change | | | | |
| Explain how these factors cause climate change | | | | |
| **Chapter 3 Climate change – human factors** | | | | |
| Describe the difference between the natural and the enhanced greenhouse effect | | | | |
| Give examples of some human activities that contribute to climate change | | | | |
| Explain how these activities affect the climate | | | | |
| **Chapter 4 Climate change – its effects** | | | | |
| Give examples of some of the ways that changes in the climate can positively affect countries | | | | |
| Give examples of some of the ways that changes in the climate can negatively affect countries | | | | |
| **Chapter 5 Climate change – coping with its effects** | | | | |
| Give examples of strategies to reduce the effects of climate change at a local level | | | | |
| Describe some strategies that can be used on a national level to reduce the effects of climate change | | | | |
| Give examples of strategies to reduce the effects of climate change on an international level | | | | |
| **Chapter 6 Climate change – case study of Bangladesh** | | | | |
| Describe the human and physical geography of Bangladesh | | | | |
| Explain some effects of climate change on Bangladesh | | | | |
| Give reasons why Bangladesh finds it difficult to deal with the impact of climate change | | | | |

| | | | | |
|---|---|---|---|---|
| **Chapter 7 Climate change – case study of Florida, USA** | | | | |
| Describe the human and physical geography of Florida | | | | |
| Explain some of the effects of climate change on Florida | | | | |
| Give reasons why Florida finds it easier to deal with the impacts of climate change | | | | |
| **Chapter 8 Structure of the Earth** | | | | |
| Explain the meaning of the term *natural (environmental) hazard* | | | | |
| Describe the three main layers of the Earth | | | | |
| Describe the two types of the Earth's crust | | | | |
| **Chapter 9 Crustal plates and plate boundaries** | | | | |
| Give a definition of a crustal plate | | | | |
| Describe the four types of plate boundary | | | | |
| Describe the activities which take place at each plate boundary | | | | |
| **Chapter 10 Volcanoes** | | | | |
| Describe the location of volcanoes around the world | | | | |
| Explain the formation of volcanoes at plate boundaries | | | | |
| Describe the features of a volcano | | | | |
| **Chapter 11 The eruption of Mt. St. Helens, 1980** | | | | |
| Explain the reason for the eruption | | | | |
| Give examples of the effects of the eruption on the landscape | | | | |
| Describe the impact of the eruption on the people | | | | |
| **Chapter 12 Managing the eruption of Mt. St. Helens, 1980** | | | | |
| Give examples of help that was given before the eruption | | | | |
| Describe some of the short-term aid needed following the eruption | | | | |
| Describe what long-term aid is and why it was needed | | | | |
| **Chapter 13 Earthquakes** | | | | |
| Describe where earthquakes occur | | | | |
| Explain how an earthquake happens | | | | |
| Describe how earthquakes are measured | | | | |

# 'I CAN DO' CHECKLIST

| | | | | |
|---|---|---|---|---|
| **Chapter 14 The cause of the Japan earthquake, 2011** | | | | |
| Explain why Japan experiences earthquakes | | | | |
| Describe the cause of the 2011 earthquake | | | | |
| Give reasons why predicting earthquakes is difficult | | | | |
| **Chapter 15 The effects and management of the Japan earthquake, 2011** | | | | |
| Describe the impact of the earthquake on the landscape | | | | |
| Give examples of how the earthquake affected the people of Japan | | | | |
| Describe how Japan coped with this earthquake | | | | |
| **Chapter 16 Tropical storms** | | | | |
| Describe what a tropical storm is | | | | |
| Describe the main features and locations of tropical storms | | | | |
| Describe the conditions needed to create a tropical storm | | | | |
| **Chapter 17 The cause of Hurricane Irma, 2017** | | | | |
| Describe the differences between tropical storms and hurricanes | | | | |
| Explain the conditions that formed Hurricane Irma | | | | |
| Describe the path of Hurricane Irma | | | | |
| **Chapter 18 The impact and management of Hurricane Irma, 2017** | | | | |
| Explain the impact of Hurricane Irma on the landscape | | | | |
| Give examples of the impact of Hurricane Irma on the people | | | | |
| Describe the relief effort after Hurricane Irma | | | | |
| **Chapter 19 The tundra climate** | | | | |
| Describe the tundra climate | | | | |
| Give examples of countries with a tundra climate | | | | |
| Interpret a climate graph for tundra regions | | | | |
| Give examples of ecosystems within tundra regions | | | | |
| **Chapter 20 How the tundra environment is used and misused** | | | | |
| Give examples of ways that people use tundra landscapes to their advantage | | | | |
| Provide examples of ways that tundra landscapes are misused | | | | |

| | | | | |
|---|---|---|---|---|
| **Chapter 21 Alaska: the effects of human activities** | | | | |
| Describe the main human activities in Alaska | | | | |
| Describe the effects that these activities are having on the people and the environment | | | | |
| **Chapter 22 The management of human activities in the tundra** | | | | |
| Describe some of the methods used to minimise the impact of human activity in the tundra | | | | |
| Give detailed examples of strategies used to minimise the impact of human activity in the tundra | | | | |
| **Chapter 23 The equatorial forest climate** | | | | |
| Describe the equatorial climate | | | | |
| Give examples of countries with an equatorial climate | | | | |
| Interpret a climate graph for equatorial regions | | | | |
| Give examples of ecosystems within equatorial regions | | | | |
| **Chapter 24 Uses of the equatorial rainforest** | | | | |
| Describe the traditional uses of the equatorial rainforests | | | | |
| Describe recent activities in the equatorial rainforests | | | | |
| **Chapter 25 The Amazon rainforest: effects of deforestation** | | | | |
| Give examples of how deforestation affects people | | | | |
| Describe the effects of deforestation on the environment | | | | |
| **Chapter 26 The management of human activities in equatorial rainforests** | | | | |
| Give examples of management strategies used in the Amazon region | | | | |
| Explain how these strategies are helping to minimise the damage caused by deforestation | | | | |
| **Chapter 27 Health in developing countries** | | | | |
| Describe the main reasons for ill-health in developing countries | | | | |
| Explain the effects of ill-health on the people and their countries | | | | |
| Describe some solutions to ill-health in developing countries | | | | |

| | | | | |
|---|---|---|---|---|
| **Chapter 28 Health in developed countries** | | | | |
| Explain why developed countries have less disease than developing countries | | | | |
| Describe the areas within developed countries which have the worst health | | | | |
| Describe the main factors affecting health in developed countries | | | | |
| **Chapter 29 Malaria – its cause and transmission** | | | | |
| Describe the distribution of malaria around the world | | | | |
| Explain the causes of malaria | | | | |
| Describe how malaria is transmitted | | | | |
| **Chapter 30 Malaria – factors in its distribution** | | | | |
| Describe the physical environment where malaria is likely to be found | | | | |
| Explain the human factors that contribute to the spread of malaria | | | | |
| Give examples of methods used to control malaria | | | | |
| **Chapter 31 Heart disease – its causes** | | | | |
| Describe the main causes of heart disease | | | | |
| Draw and interpret graphical data | | | | |
| **Chapter 32 Heart disease – methods of control** | | | | |
| Describe the factors in the distribution of heart disease | | | | |
| Give examples of methods of controlling heart disease | | | | |
| Explain the role played by the National Health Service in preventing and treating heart disease | | | | |
| **Chapter 33 HIV/AIDS – its distribution** | | | | |
| Explain how HIV/AIDS affects people | | | | |
| Describe the global distribution of HIV/AIDS | | | | |
| Draw and interpret choropleth maps | | | | |
| **Chapter 34 HIV/AIDS – causes, effects and treatment** | | | | |
| Describe the factors affecting the distribution of HIV/AIDS | | | | |
| Explain the consequences of HIV/AIDS for a country | | | | |
| Give examples of how HIV/AIDS can be treated | | | | |

# Index

affluence 173
afforestation 141
Africa, and HIV/AIDS 177–9, 180–1
aid 64–6, 83, 99
AIDS (acquired immunodeficiency syndrome) 177–9, 180–1, 186, 188
air temperatures 6
Alaska 115–8, 121–3, 146
*Anopheles* mosquitoes 159, 162
anti-malarial drugs 163

Bangladesh, case study 26–30
Barbuda 96–98, 102, 104–6
biodiversity changes 3
Biodiversity Action Plans 122

carbon dioxide 14, 15, 138
case studies
    Bangladesh 26–30
    Florida 32–5
    South Africa 182
CFCs (chlorofluorocarbons) 14, 16
climate change
    case studies
        Bangladesh 26–30
        Florida 32–5
    causes 3, 37, 138
    dealing with 22–4, 35, 40, 122
    effects of 19–21, 35, 38, 40, 42
    human causes 13–14, 38, 138
    human factors 13–16
    meaning 19
    measuring 3–4
    physical factors 7–10
    policies 23–4
    as a term 2
coal 14, 15
collision plate boundaries 48–9
communications 117
conservative plate boundaries 48, 69, 72
constructive plate boundaries 48, 53–6, 70
core 44–5
crust 44–5
crustal plate boundaries 53, 68–71
crustal plates 47–9, 53
Cuba 97
cyclones 88

deforestation 14, 129, 132–3, 136–8, 140–1
destructive plate boundaries 47–8, 49, 53, 54, 70, 76
developed countries
    health 148–9, 153–6
developing countries
    health 148–9
diet 155–6, 167, 172–3, 174
Doha Amendment 25
drug abuse 181

Earth, stretch, tilt and wobble 8
earthquakes 67–71, 74–7, 79–83, 103
Earth's structure 44–5
ECCP (European Climate Change Programme) 24

enhanced greenhouse effect 13–14
environmental hazards, Earth's structure 44–5
equatorial regions 125–9, 131–3, 136–8, 140–1
    climate 12–7, 145
EU Emissions Trading System 25
European Climate Change Programme (ECCP) 24
exercise 167, 168
exports 143, 146
eye-walls 86
eyes (of storms) 86
Exxon Valdez oil spill 118, 146

Fiji 42
Florida, case study 32–5
fossil fuels 14, 15, 38, 41, 45
Fukushima nuclear power plant 80–1

glaciers, and climate change 4
global warming 19, 118, 122
greenhouse effect 1–2, 13–14
greenhouse gases 1–2
health
    comparisons between countries 148–9
    developed countries 153–6
    ill-health 149–50

health care 154, 173
health education 154, 173–4
health facilities 156
heart disease 166–9, 172–4, 184
heli-logging 141
HIV/AIDS 177–9, 180–2, 186, 188
hunting 111, 113
Hurricane Irma 90–2, 96–9
hurricanes 34, 35, 85, 88, 90–2, 96–9, 101

Ice Age 8
ice core analysis 3
ice extent 4
ill-health 149–50, 154–6
indigenous peoples
    tundra 111, 116, 117, 146
    rainforest 131–2, 136–8
infrastructure 117
inheritance, and heart disease 167

Japan earthquake 74–7, 79–83

Kyoto protocol 24, 122

L waves 68
life expectancy 148
lifestyle, and heart disease 172
Little Ice Age 7
long-term aid 64

malaria 158–60, 162–3
Malaysia 143
mantle 44–5
medical care 173
methane 14, 15
mining 112, 113, 116, 117, 133, 146

mosquitoes 159, 162–4
Mt. St. Helens eruption 58–62, 64–6

National Health Service (NHS) 172–3
natural gas extraction 112–13, 114
New Zealand 103, 105
NHS (National Health Service) 172–3
nitrous oxides 14, 15–16
North Atlantic Drift 8
nuclear power plants 80–1

ocean currents 8, 9
oil extraction 112–3, 114, 116
oil spills 118
    *see also* Exxon Valdez oil spill
official aid 64, 99

P waves 68
permafrost 108, 121
plate boundaries 47–49, 70–1, 76
plate tectonics 10
pollution 113, 114, 117, 118, 155

rainforest layers 127–8
relief efforts 64–6, 83
Richter scale 68
Russia 113, 122

S waves 68
sea temperatures 4
shifting cultivation 131–2
shock waves 68
short-term aid 64
smoking 167, 174
social habits 155
solar radiation 7
South Africa, HIV/AIDS case study 182
South Korea 187
St Martin 97–8
sustainable forestry 140–1

tectonic plates 45
temperature change 2–4, 14, 39, 40, 42
Think Global. Act Local 122
tourism 113
Trans-Alaska Pipeline 113, 115–16, 117, 118, 121–2
tree ring analysis 3
tropical storms 85–8
tsunami waves 74, 76
tundra regions 106–9, 111–14, 115–18, 121–2
    climate 107–8
typhoons 88, 104

United Nations Framework Convention on Climate Change (UNFCCC) 24

vehicles 14, 117
volcanic eruptions 8–9, 54, 60–4
volcanoes 54–7
voluntary aid 64

war, and HIV/AIDS 181
WWF (World Wide Fund for Nature) 122